A TEXT BOOK OF GEOLOGY

A Text Book of
Geology

With Special Reference to India

Girija Bhushan Mahapatra MSc
Former Geologist (GSI)

CBSPD

CBS Publishers & Distributors Pvt Ltd

New Delhi • Bengaluru • Chennai • Kochi • Kolkata • Lucknow • Mumbai
Hyderabad • Jharkhand • Nagpur • Patna • Pune • Uttarakhand

A Text Book of
Geology

Copyright © Author and Publisher

ISBN-13: 978-81-239-0013-1
ISBN-10: 81-239-0013-9

First Edition: 1987
Reprint: 1989, 1990, 1994, 1997, 1999, 2000, 2001, 2002, 2003, 2004, 2005, 2006, 2007, 2008, 2009, 2010, 2011, 2012, 2013, 2014, 2015, 2016, 2017, 2018, 2019, 2020, 2022, **2024**

Published by **Satish Kumar Jain** and produced by **Varun Jain** for

CBS Publishers & Distributors Pvt Ltd
4819/XI Prahlad Street, 24 Ansari Road, Daryaganj, New Delhi 110 002, India.
Ph: 011-23266838, 23289259   Website: www.cbspd.com
                                 e-mail: delhi@cbspd.com

Corporate Office: 204 FIE, Industrial Area, Patparganj, Delhi 110 092
Ph: 011-4934 4934   Fax: 011-4934 4935
                        e-mail: publishing@cbspd.com; publicity@cbspd.com

Branches

- **Bengaluru:** Seema House 2975, 17th Cross, KR Road, Banasankari 2nd Stage, Bengaluru 560 070, Karnataka, India
  Ph: +91-80-26771678/79   Fax: +91-80-26771680   e-mail: bangalore@cbspd.com
- **Chennai:** 7, Subbaraya Street, Shenoy Nagar, Chennai 600 030, Tamil Nadu, India
  Ph: +91-44-26680620, 26681266   Fax: +91-44-42032115   e-mail: chennai@cbspd.com
- **Kochi:** 42/1325, 1326, Power House Road, Opp KSEB, Power House, Ernakulum Kochi 682 018, Kerala, India
  Ph: +91-484-4059061-65,67   Fax: +91-484-4059065   e-mail: kochi@cbspd.com
- **Kolkata:** 147, Hind Ceramics Compound, 1st Floor, Nilgunj Road, Belghoria, Kolkata-700056, West Bengal, India
  Ph: +033-25633055, 033-25633056   e-mail: kolkata@cbspd.com
- **Lucknow:** Basement, Khushnuma Complex, 7 Meerabai Marg (Behind Jawahar Bhawan), Lucknow-226001, UP, India
  Ph: +0522-4000032   e-mail: tiwari.lucknow@cbspd.com
- **Mumbai:** PWD Shed, Gala no 25/26, Ramchandra Bhatt Marg, Next to JJ Hospital Gate no. 2, Opp. Union Bank of India, Noorbaug, Mumbai-400009, Maharashtra, India
  Ph: 022-66661880/89   e-mail: mumbai@cbspd.com

Representatives

• Hyderabad	0-9885175004	• Jharkhand	0-9811541605	• Nagpur	0-8692091830
• Patna	0-9334159340	• Pune	0-9664372571	• Uttarakhand	0-9716462459

Printed at SRK Graphics, Delhi (India)

Dedicated to :

The Sweet Memory of my

Beloved Father—

Late Shri R. Ch. Mahapatra

Who was a source of

Perennial Inspiration to me.

—AUTHOR

PREFACE

My primary objective in preparing this book "A Text-Book of Geology" has been to present the subject matter in such a way that a student will not only find it useful from the examination point of view, but will also be able to apprehend fully the basic principles of Geology.

It is written, covering the standard syllabi of undergraduate classes of Indian Universities. Besides, earnest efforts have also been made to cover-up the whole geology syllabus of IAS and Allied Services Exam. (Preliminary) and a majority of the chapters of the syllabus of Geologists' Examination, as has been prescribed by the Union Public Service Commission.

The author who had been successful in the All India Competitive Examinations like Geologists Examinations and Civil Services Examinations in securing good positions in the merit list, has kept in view the usual difficulties encountered by the students in making a systematic approach to the subjects, while writing this book. Hope this book will be able to avoid those difficulties and the book can be used as a standard text book.

Suggestions are welcome from the readers with regards to the manner of approach, explanations and errors, if any, in this book so as to enable the author in bringing about an improved version of the book in the subsequent editions.

The whole credit for writing this book goes to my father whose inspiration and encouragements were solely responsible for my being the topper in the University in my B.Sc. Exams.

I am very much thankful to a number of my friends, whose goodwill and co-operation was of much help to me in completing this book ; name of Sh. G. Panigrahi, Sh. S.C. Samal and Sh. Shashi Bhushan need special mention here in this context.

I am thankful to the 'CBS Publishers' for making best efforts to bring out this book in a presentable form.

—AUTHOR

CONTENTS

PART VI

ECONOMIC GEOLOGY

PART VII

ORIGIN, MODE OF OCCURRENCE AND DISTRIBUTION (In India) AND ECONOMIC USES

PART VIII

—STRATIGRAPHY   —PALAEONTOLOGY

BIBLIOGRAPHY

1. *A Text Book of Mineralogy.*
   —E.S. Dana and W.E. Ford, Asia Publishing House.

2. *Dana's Manual of Mineralogy.*
   —Hurtburt Cornelius (Jr.), John Wiley and Sons.

3. *Mineralogy.*
   —E.H. Kraus, W.F. Hunt and L.S. Ramsdell, Mc-Graw Hill Book Co.

4. *Crsytallography.*
   —Wades and Mattox, Oxford and IBH Publishing House Co.

5. *Kemp's and Hand Book of Rocks.*
   —F.F. Grout, D. Van Nostrand Company.

6. *Petrography and Petrology.*
   —F.F. Grout, Mc-Graw Hill Book Co.

7. *Igneous and Metamorphic Petrology.*
   —F.J. Turner and J. Verhoogen, Mc-Graw Hill.

8. *Igneous Rocks and Minerals.*
   —E.E. Wahlstrom, John Wiley and Sons.

9. *Evolution of the Igneous Rocks.*
   —N.L. Bowen, Princetan University Press.

10. *Principles of Petrology.*
    —G.N. Tyrrel, Methuin and Co. Ltd.

11. *Petrography.*
    —H. Willians, F.J. Turner and C.M. Gilbert, Freeman Co.

12. *Basic Concepts of Physical Geology.*
    —E.W. Spencer, Oxford and IBH Publishing Co.

13. *Elements of Structural Geology.*
    —E.S. Hill, Methuin

14. *Structural Geology.*
    —Marland P. Billings, Prentice Hall.

15. *Field Methods*
    —F.H. Lahee, John Wiley and Sons.

PART—I

PHYSICAL GEOLOGY

CHAPTER 1

ORIGIN OF THE EARTH

The Earth is a member of the planetary system of the sun. The principal theories which have been advanced to explain the origin of the earth, can be divided into two groups. All these theories have in common the idea that the planets evolved from the sun. They differ as to the manner in which it occurred.

1. In one group, material is pulled out of the sun by an external force such as gravitational pull resulting from the dynamic encounter or near-collision of the sun with another star. These are also known as catastrophic theories.

2. The second group holds that the planets became isolated masses of matter as the material of the solar-system condensed into the sun. These are also known as natural or evolutionary theories.

The earlist theory for the origin of the earth was put forward by Kant and Laplace (in 1796). It is known as Nebular Hypothesis.

(a) **Nebular hypothesis.** According to this hypothesis, the solar system evolved from a single, large, flat, rotating nebula that extended beyond the position of the most distant planet. As this nebula contracted the mass became increasingly concentrated towards the centre and in doing so it would have rotated more rapidly to conserve angular momentum, until (at some period during the contraction) the speed of the outermost rim of the disc would have become sufficient for the centrifugal force to be as great as the inward gravitational attraction. At this position a ring of matter was left while the contraction of the remaining matter continued. In this manner successive rings of matter were left behind the contracting mass. Subsequently the material within each ring was drawn together and planets and satellites were formed.

This theory was rejected when it was learnt that the angular momentum of the solar system is concentrated in the planets and not in the sun. This is also not compatible with the idea that the mass of matter rotated more rapidly as it condensed.

(*b*) In 1944, a German physicist, C.F. Von Weizsacker proposed a modification of the nebular hypothesis.

(*c*) **Proto-planet hypothesis of Kuiper.** It is the most popular hypothesis within the past few years. It visualises a slightly flattened and slowly rotating disc-shaped solar nebula bulging out from the equator of the sun. In composition this cloud is similar to the sun and contains mostly hydrogen and helium with small amounts of heavier elements, but the disc and the sun itself are thought of as being cool. The disc containing about one tenth the mass of the sun finally becomes internally unstable and breaks up into smaller concentrations called protoplanets. Within each protoplanet, the heavier elements tend to settle towards the centre and the lighter particles and gases remain in outer shells.

(*d*) **Bi-parental origin of the earth.** This theory belongs to the group of dynamic-encounter theories and put-forward by Chamberlin and Moulton, "as another star approached the sun, tremendous tides were set up on the surface of the sun and these tides or filaments of hot gases were pulled out from the sun. As the star passed, these arms of gas were given a rotational motion. After the star was gone, the gaseous matter in these arms condensed into solid material and gradually drew together to form planets.

Similar hypothesis indicating biparental origin of the earth were propounded by Jeans and Jeffry and also by Lyttleton. But these theories gradually lost its wide acceptance because of the absence of relevant evidences to support it.

Now-a-days, the Big-Bang theory as is being propounded by the American astronomers, that the universe, solar system etc. are the result of an explosion within the nebula, is getting better acceptance :

CHAPTER 2

AGE OF THE EARTH

Even though there is a divergence of opinion among the propounders of the various theories on 'the origin of earth' ; it is almost unanimously believed that earth took its birth in a hot, gaseous, molten state. Different criteria have been used and various factors have been used to determine the age of the earth. The two distinct processes of estimation are as follows :

1. Dating of Geological-formations and geological-events by indirect methods (relative age).

2. Radio-active methods for determining the actual age.

1. Indirect methods. It includes various processes of determining the age, like varve-clocks, sedimentary-clocks, salinity-clock, etc. Besides, attempts were made to determine the age of the earth from the rate of cooling of the earth, from the evolutionary changes of animals etc.

(*a*) **Varve-clock method.** It uses the fine sedimentary deposits of glacial origin, which represents the annual accumulation of the layers marked by variation in colour and gradation in the size of materials constituting the layers.

Geologic time ranging from '0' to 10,000 years only could be counted on this varve clock, where varve occurs as in Kashmir Himalayas.

(*b*) **Sedimentation-clock.** In this case the average annual rate of deposition of sediments and the thickness of all strata deposited during the whole geological history, are taken into account. Although this method is full of imperfections and variations from place to place, the age of the cambrian as determined by this and other accurate methods closely approach each other.

Here it is accepted that there is an average rate of deposition of about 1 ft. per 755 years ; thus the beginning of the Cambrian-sedimentation comes to about 510 million years.

(*c*) **Salinity-clock.** All the salts of the seas of today have been acquired from the land by its weathering and erosion through ages.   By determining the yearly rate of increase of salinity the age of the earth was calculated to be 100 million years by Joly.

(*d*) **Rate of cooling of the earth.**   Assuming the initial temperature of the earth to be 3900°C, Lord Kelvin determined the age of the earth to be between 20 to 400 million years.   But, it was full of imperfections, since the generation of radio-active heat and other allied factors were not taken into account.

(*e*) **Evolutionary changes of animals.**   As we know, the first-formed animals were unicellular which underwent various phases of the evolutionary processes and multi-cellular organisms with more complexities came into being gradually and man is considered to be the most evolved one.   Taking into account the evolutionary developments, Biologists determined the age of earth to be 1000 million years.

2. **Radioactive methods.**   The basic principle underlying all the radioactive methods is that "a radioactive parent element decays into a stable daughter element at a constant rate".   For geological purposes the unit of time is one year.   Usually the "Half-life" period is determined and accordingly it is equated to find-out the age of the earth.   The followings are the common methods which are used for the purpose of determining the age of earth :

(*a*) **Uranium-lead method.**   Here two isotopes of uranium can be used, U^{238} and U^{235} ; the half-lives of which are 4498 million years and 713 million years respectively.

(*b*) **Thorium-lead method.**   Thorium (232) yields lead (208) through radio-active decay and the half-life of thorium is 13,900 million years.

(*c*) **Potassium-argon method.**   Potassium is having three isotopes like 39 K, 40 K and 41 K. But only 40 K is radio-active. The half-life of 40 K to convert itself into 40-Ar is 11,850 million years.

(*d*) **Rubidium-strontium method.**   Rubidium (Rb^{87}) yields strontium (Sr^{87}) and its half-life period is about 50,000 million years.

It is a particularly valuable method for metamorphic rocks and for pre-cambrian rocks.

(*e*) **Radio-carbon method.**   The carbon (C^{14}) is radio-active, with a half-life of 5570 years. All organisms take in C^{14} and a constant level of this isotope is maintained by all living organisms. At death the C^{14} intake ceases and its proportion decreases at constant rate.   This method is especially useful for dating relatively recent materials up to 70,000 years.

The application of radio-active methods for the determination of the age of earth, has been successful in the estimation and the age of the earth, accordingly, comes to about 4500 million years, *i.e* 4·5 billion years.

CHAPTER 3

STRUCTURE OF THE EARTH

The study of the structure of the earth focusses on its layered structure, and the variations in the density and temperature, at various depths. The shape of the earth is that of a spheriod with mean equatorial radius of 6378-388 km. and polar radius of 6356·912 km. Concisely speaking, the earth is a globe having a radius of 6371 kilometres.

Direct observation of the interior of the earth is not possible as the interior becomes hotter with depth which is convincingly indicated by the volcanic eruptions. Apart from the seismological studies, other important sources of data, even though indirect, logically prove that the earth's body comprises several layers, which are like shells resting one above the other. These layers are distinguished by their physical and chemical properties, particularly, their thickness, depth, density, temperature, metallic content and rocks.

The layered structure of the earth-developed during the process of its transformation from a hot-gaseous state to the present state. During these processes, the heavier material sank down and the lighter material floated up and consequently because of the differential densities of the materials constituting the earth, they got separated and formed layers of different densities.

Broadly, the earth's interior has been divided into three major parts :

1. The Crust. 2. The Mantle. 3. The Core.

Inferences obtained through seismological studies. The seismic waves which result during the occurrence of earthquakes are chiefly of three types as .

(*a*) Primary waves (P-waves).

(*b*) Secondary or Sheer Waves (S-waves).

(*c*) Rayleigh (R) waves which are also known as 'L'-waves.

These seismic waves differ from each other in respect of their propogation velocity, wave-length and path of travel. Their nature of vibration is also different as some of them have longitudinal-vibration, others have vibrations of transverse-nature.

From several seismological studies, it has been inferred that

(*i*) **The** three segments of the earth *i.e.*, Crust, Mantle and Core, are separated by two sharp breaks, which are usually known as major discontinuities.

(*ii*) The crust is having a thickness of about 33 kms.

(*iii*) The crust is composed of heterogeneous rocks.

(*iv*) The second major segment of the earth *i e.*, the mantle extends from below the crust to a depth of 2900 kms.

(*v*) The third major segment of the earth extends from below the mantle up to the centre of the earth and is known as 'the core'.

Apart from the aforesaid facts, it also gives evidences regarding the existence of a number of minor discontinuities within the earth, which may be because of

(*a*) Changes in the chemical composition of the materials.

(*b*) Changes in the density of the materials.

(*c*) Changes in the state of a given material *i.e.*, whether it is in solid, liquid or viscous state.

(*d*) Changes in the physical properties of minerals, etc.

1. The crust. It is the top-most layer of the earth. Its thickness over the oceanic areas is generally 5 to 10 kms, whereas on the continental area, it is 35 kms and the thickness ranges from 55 to 70 kms in orogenic belts.

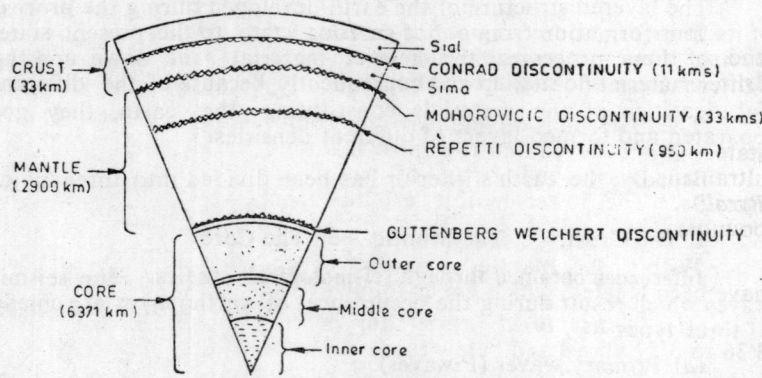

Fig 3·1. Divisions of the Earth's Interior.

Sub-divisions of the crust. The crust of the earth is sub-divided into sub-layers as follows :

(*i*) Sial (*ii*) Sima

(*i*) **Sial.** It is also known as the Upper-Continental-Crust. It consists of all types of rocks like igneous, sedimentary and metamorphic, which are exposed at the land-surface. It is rich in silica and aluminium.

Its composition is usually granitic to grano-dioritic. In the ocean-basins, they are floored by a basaltic-horizon which are poorer in potassium and richer in aluminium than the basalts of the land-surface and are called 'Oceanic-tholeiites'.

Conrad Discontinuity separates the sial-layer from the under-lying sima-layer. This discontinuity is a second-order discontinuity and is located at a depth of 11 kms.

(*ii*) **Sima.** It is also known as Lower-Continental crust. Its thickness is about 22 kms. It includes two parts :

(*a*) Outer Sima

(*b*) Inner Sima.

Together they are basaltic in composition and this layer is rich in silica and magnesium.

(*a*) Outer sima, extends up to a depth of 19 kms and comprises rocks of intermediate composition.

(*b*) Inner sima is located at a depth of 19 kms and extends up to 33 kms. It is of basic to ultra-basic in composition.

2. Mantle. It is separated from the over-lying crust by the Mohorovicic-Discontinuity which is a first-order discontinuity. Its thickness is about 2865 kms. If forms 83 % of the earth by volume and 68 % by mass.

It is the source-region of most of the earth's internal energy and of forces responsible for ocean-floor spreading, continental drift, orogeny and major earthquakes.

The material is olivine-pyroxene complex, which exists in a solid state. It is believed that the upper mantle has a mix of 3-parts of ultramafic rocks and 1-part of basalt and the mix is known as *Pyrolite*. The lower mantle extends from 1000 km to the core-boundary.

Within the mantle, a number of second-order discontinuities have been located, which are as follows :

(*i*) *Density break* at 80 km depth ; density changes from 3·36 to 3·87 above and below the level, respectively.

(*ii*) *Gravity break* at 150 km depth ; gravity changes from 984 cm/sec^2 to 974 cm/sec^2 till it reaches at a depth of 1200 kms.

(*iii*) At 700 km-depth, there changes the capability of the materials in storing amount of elastic-strain energy. Up to 700 kms the capability is more.

(*iv*) *Repetiti discontinuity.* At 950 km depth. It marks the lower limit of the very rapid rise in the velocity of seismic vibrations.

(*v*) *Gravity-break.* At 1200 km depth, gravity attains its minimum value *i.e.*, 974 cm/sec^2, thereafter it rises upto 1068 cm/sec^2 at the core-boundary.

3. The core. It is separated from the mantle by the Guttenberg Weichert Discontinuity and extends up to the centre of the earth. It consists of three parts :

(*i*) Outer-core,   (*ii*) Middle-core, and (*iii*) Inner-core.

(*i*) *Outer core.* It extends from 2900 to 4982 kms. It is considered to be in a state of homogeneous fluid and it does not transmit S-Waves.

(*ii*) *Middle core.* It is a transition layer, extends from 4982 kms to 5121 kms. The material is in a fluid to semi-fluid state.

(*iii*) *Inner core.* It is believed to contain metallic nickel and iron and is called 'nife'. It is probably solid with a density of about 18. Its thickness is 1250 kms.

Other Important Facts

1. The central-temperature is estimated to be 6000°C.

2. The central-pressure is 392×10^8 bars in C.G.S unit.

3. Density at the centre is 18 gm/cm^3.

4. Lithosphere constitutes the upper horizon of the crust only up to a depth of about 16 km.

5 Asthenosphere is the layer beneath the lithosphere which virtually has no strength to resist deformation.

CHAPTER 4

ATMOSPHERE, HYDROSPHERE, LITHOSPHERE AND THEIR CONSTITUENTS

1. Atmosphere. The air which envelopes the earth and extends up to a considerable height from the surface of the earth is called the atmosphere. It consists of a mixture of various gases and is held to the earth by gravitational attraction. This envelope of air is densest at sea level and thins rapidly upwards. The atmosphere constitutes a very insignificant percentage of the mass of the earth.

Structure of the atmosphere. The structure of the atmosphere consists of five basic layers :

(*a*) the troposphere,

(*b*) the stratosphere,

(*c*) the mesosphere,

(*d*) the ionosphere, and

(*e*) the magnetosphere and exosphere.

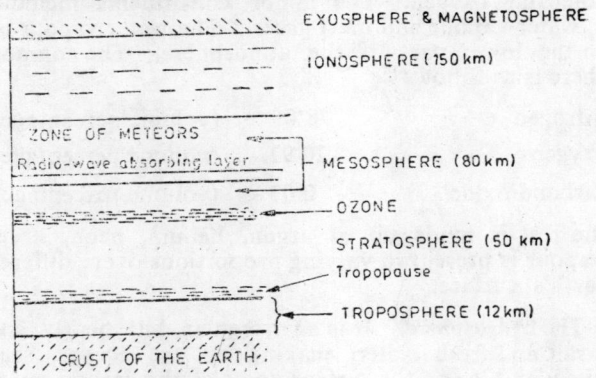

EXOSPHERE & MAGNETOSPHERE

IONOSPHERE (150 km)

ZONE OF METEORS
Radio-wave absorbing layer — MESOSPHERE (80 km)

OZONE

STRATOSPHERE (50 km)
Tropopause

TROPOSPHERE (12 km)

CRUST OF THE EARTH

Fig. 4·1.

(*a*) **Troposphere.** It extends up to a height of 12 km on an average, from the surface of the earth. At the equator the thickness of the troposphere is the greatest. It is densest of all the layers. It is the locale of all the vital atmospheric processes which create the climatic and weather conditions on the earth's surface. About half the mass of air comprising the entire atmosphere is concentrated in this zone.

The temperature of air in the troposphere decreases at the rate of 1°C per 165 metres of height.

Tropopause is an undefined region lying between troposphere and stratosphere, and there prevails a constant temperature in this region.

(*b*) **Stratosphere.** Here air is at rest. It is an isothermal region and is free of clouds, dust and water vapour. It extends up to a height of about 50 kms. Its upper strata is rich in ozone which prevents ultraviolet radiation by absorbing them and a very little is filtered through it, which does not harm living organisms.

(*c*) **Mesosphere.** It is a very cold region and extends up to a height of 80 km from the surface of the earth. At a height of 60 kms there is an intermediate layer which is known as radio-waves absorbing layer.

(*d*) **Ionosphere.** It is a region of electrically charged or ionised air lying next to mesosphere. It protects us from falling meteorites. It extends up to a height of 150 kms.

(*e*) **Exosphere.** It is the uppermost region of ionosphere and is the fringe of atmosphere. Its boundaries are not known.

Composition of atmosphere. The atmospheric composition of the troposphere reveals two major constituents : molecular nitrogen and molecular oxygen. The minor constituents include carbon dioxide, water vapour and inert gases. The water vapour is concentrated in the lower part of the atmosphere. The composition of atmosphere is as follows :

Nitrogen	78·09	(volume percentage)
Oxygen	20·97	(volume percentage)
Carbondioxide	0·03	(volume percentage)

The rest is made up of argon, helium, neon, krypton etc. Water vapour is present in varying proportions over different parts of the earth's surface.

2. The hydrosphere. It is an irregular but nearly continuous layer of salt and fresh water, making up the oceans, seas, lakes, rivers, snow and ice on the surface and also the ground water.

The world ocean covers about 71% of the global surface and its average depth is about 3800 metres.

Composition of sea-water. Sea-water is a solution of salts—a brine—whose ingredients have maintained approximately fixed proportions over a considerable span of geologic time. Of the various elements combined in these salts chlorine alone makes up 55% by weight of all the dissolved matter and sodium makes up 31%. Magnesium, calcium, sulphur and potassium are the other four major elements in these salts. Sea-water also holds in solution small amounts of all the gases of the atmosphere.

Common elements present in the hydro-sphere-ocean water, as shown by Brian Mason is as follows :

Oxygen	85·79%	(by weight)
Hydrogen	10·67%	(by weight)
Chlorine	1·898%	(by weight)
Sodium	1·056%	(by weight)
Magnesium	0·127%	(by weight)
Sulphur	0·088%	(by weight)
Calcium	0·040%	(by weight)
Potassium	0·038%	(by weight)
Bromine	0·007%	(by weight)
Carbon (inorganic)	0·003%	(by weight)
Strontium	0·001%	(by weight)
	99·718	

The rest is made up by other dissolved gases.

3. **The lithosphere.** It is the general term for the entire solid earth realm, *i.e.*, the crust. According to the recent concept, the term lithosphere is used for the crust and upper part of mantle which is considered to be elastically very strong. The lithosphere is underlain by asthenosphere.

As per the estimation, made by Clarke and Washington, the lithosphere consists of 95% igneous rocks, 4% shale and 0·75% sandstone and 0·25% limestone (the metamorphic rocks being altered equivalents of one or other of these rocks).

Chemical composition of the lithosphere. It has been observed that while the continents are distinctly sialic the ocean floor, represent the upper sima, are distinctly basic in nature. The following is the composition in terms of elements (by weight), as given by Clarke and Washington :

Oxygen	46·71	Sodium	2·75
Silicon	27·69	Potassium	2·08
Aluminium	8·07	Magnesium	2·60
Iron	5·05	Titanium	0·62
Calcium	3·65	Hydrogen	0·14
		Remaining elements	0·64
			100·00

In terms of oxides (by weight) the chemical composition of the lithosphere is as follows :

SiO$_2$	59·07	Silicon oxide
Al$_2$O$_3$	15·22	Aluminium oxide
CaO	5·10	Calcium oxide
FeO	3·71	Ferrous (iron) oxide
Na$_2$O	3·71	Sodium oxide
MgO	3·45	Magnesium oxide
K$_2$O	3·11	Potassium oxide
Fe$_2$O$_3$	3·10	Ferric (iron) oxide
H$_2$O	1·30	Water
TiO$_2$	1·03	Titanium oxide
Remaining oxides	1·20	
	100·00	

From the above, it is evident that 99% of the upper crust is made up of only 10 elements, with oxygen accounting for slightly less than 50% Besides, the above mentioned ten oxides constitute more than 98% of the lithosphere, with silica being the most abundant.

The mineralogical composition of the lithosphere shows that plagioclase feldspar is the most abundant mineral and apatite is the least one. The estimate, as given by Clarke and Washington is as follows :

(*i*) Quartz	11% by volume
(*ii*) Alkali feldspar	16% by volume
(*iii*) Plagioclase feldspar (Andesine)	47%
(*iv*) Amphiboles and Biotite	20%
(*v*) Magnetite	5%
(*vi*) Apatite	1%
	100

The above analysis gives only the average mineralogical composition of the lithosphere, but it does not in any way represent the composition of the earth as a whole or even the crust as a whole.

CHAPTER 5

PROCESSES OF WEATHERING

Introduction. Weathering is the general term applied to the combined action of all processes causing rocks to be disintegrated physically and decomposed chemically because of exposure at or near the earth's surface.

In particular, weathering occurs, where rocks and minerals come in contact with the atmosphere, surficial water, and organic life under conditions that are normal to the surface of the earth.

Weathering is the initial stage in the process of denudation. An essential feature of the process is that it affects rocks *in situ* ; no transportation is involved. The products of rock-weathering tend to accumulate in a soft surface layer, called *regolith*. The regolith grades downward into solid, unaltered rock, known simply as *bedrock*.

Weathering helps erosion, but is not a part of it. There can be weathering without erosion and erosion without weathering.

Types of weathering. There are three main types of weathering :

1. Physical weathering.
2. Chemical weathering.
3. Biological weathering.

Factors affecting weathering. All the processes of weathering are affected by rock structure, climate, topography and vegetation.

Rock structure refers to mineralogical composition, joints bedding planes, faults, fractures, pores and its integral hardness. The degree of weathering of the source area, *i.e.*, the area where weathering operates is controlled by the nature of the pre-existing rocks to a greater extent.

Climate is the sum-total of the meteorological elements like temperature, moisture, including both humidity and precipitation, wind, air-pressure and evaporation. Climate determines whether physical or chemical weathering will predominate and the speed with which these processes will operate.

Topography directly affects weathering by exposing rocks and indirectly through the amount of precipitation, temperature and vegetation.

Surfaces covered with vegetation are protected and bare surfaces are weathered to a greater extent.

Agents of weathering. The principal agents of the transformation of rocks in the mantle of waste are water, oxygen, carbondioxide, acids, organisms, and variations of temperature. These agents affect equally good in case of both physical and chemical processes of weathering.

1. Physical weathering processes. This process refers to the mechanical disintegration of rocks in which their mineralogical composition is not changed. This is brought about chiefly by temperature changes, *e.g.*, thermal expansion and contraction. The followings are some of the important processes of physical weathering.

(*a*) **Exfoliation.** In this case thin sheets of rock split off owing to differential expansion and contraction during heating and cooling over the diurnal temperature range.

Sometimes, it is the result of *unloading* in which case, because of the removal of the overlying rocks, the pressure on the igneous rocks beneath them is also diminished and this results in the expansion of igneous rocks and in the formation of large scale fractures parallel to the surface topography. Sheets between the fractures are detached from the main mass which thus suffers fragmentation.

(*b*) **Crystal growth.** The soluble constituents of the rocks or minerals, enter the rocks through fractures and joints, along with water. With the evaporation of water the solution is precipitated to form crystals or crystalline aggregates and as they grow, they exert large expansive stresses, which help in breaking up some rocks.

(*c*) **Freezing of water.** Water, as we know, expands by about 9·05 percent in volume when it freezes. The water seeps down into the fracture and under suitable climatic condition, begins to freeze at the top of the fracture first. As freezing continues, the pressure exerted on the walls becomes more and more intense. which results in widening the existing fracture and new fractures form. This is the dominant mode of weathering, in climates where there is repeated freezing and thawing. This is also known as *Frost action*.

(*d*) **Differential expansion** Rock-forming minerals expand when heated, but contract when cooled. Where rock surfaces are exposed daily to intense heating by direct solar rays, alternating with intense cooling by longwave radiation at night, the resulting expansion and contraction of mineral-grains tends to break-them apart.

The intense heat of forest and bush fires is known to cause rapid flaking and scaling of exposed rock-surfaces.

2. Chemical processes of weathering. It is also known as mineral alteration, consists of a number of chemical reactions, all these reactions change the original silicate minerals of igneous rock, *the primary minerals*, into new compounds, the *secondary minerals*, that are stable in the surface environment. Besides, sedimentary and metamorphic rocks are also substantially affected by the chemical processes of weathering. Chemical weathering is more important than mechanical weathering in almost all the climatic regions.

The atmosphere contains a number of constituents that can react with minerals. Most important of these are water, carbon-dioxide, and oxygen. The effectiveness of these chemical constituents depends on the composition of the rocks and size of the particles that make them up. For example, Quartz is a very stable substance, so rocks composed primarily of quartz decompose very slowly ; whereas the ferromagnesian minerals are highly susceptible to chemical weathering.

Three processes are notably responsible for chemical weathering :

(*a*) Oxidation.

(*b*) Hydration.

(*c*) Carbonation.

(*a*) **Oxidation.** The presence of dissolved oxygen in water in contact with mineral surfaces leads to oxidation ; which is the chemical union of oxygen atoms with atoms of other metallic elements. Oxygen has a particular affinity for iron compounds and these are among the most commonly oxidised materials.

(*b*) **Hydration.** The chemical union of water with a mineral is called hydration. It is sometimes confused with 'hydrolysis', the reaction between water and a compound. The process of hydration is particularly effective on some aluminium bearing minerals, such as feldspar.

(*c*) **Carbonation.** Carbon-dioxide is a gas and is a common constituent of the earth's atmosphere. Rain water in course of its passage through the atmosphere, dissolves some of the carbon-dioxide present in the air. It thus turns into a weak acid called carbonic acid, H_2CO_3, and is the most common solvent acting on the crust. The effect of this process is well noticed in the limestone or chalk areas in the humid regions of the world.

Besides the above, another process known as *"Solution"* is quite significant in bringing about the chemical weathering of rocks. In this case, some of the minerals get dissolved by water and thus removed in solution, for example gypsum, halite etc,

3. Biological-weathering. This process of weathering is mainly related to the activities of various organisms. Organisms, mainly plants and bacteria, take part in the transformation of rocks at the surface, in the following ways :

(*a*) Bio-physical processes.

(*b*) Bio-chemical processes.

'(*a*) **Bio-physical processes.** (*i*) Plant-roots, growing between joint blocks and along minute fractures between mineral grains, exert an expansive force tending to widen those openings and sometimes create new fractures.

(*ii*) Insects like earth-worm, snail etc. loosen the soil cover and create suitable conditions for the various external agencies to have their own action on the underlying rocks, which ultimately lead to rock weathering.

(*b*) **Bio-chemical processes of weathering.** (*i*) Sometimes, certain groups of bacteria, algae and mosses break rock-forming silicates down directly, removing from them elements like silicon, potassium, phosphorous, calcium, magnesium, that they need as nutrients. This transformation occasionally occurs on a large scale and is decisive in the alteration of parent rocks and facilitate rock weathering.

(*ii*) After the death of animals or plants, with their subsequent decay and degeneration, chemically active substances are produced, which are capable of bringing about rock-weathering. For example, humic acid which is formed during decay and degeneration of plant life is capable of bringing about rock weathering effectively, to some extent.

Thus, in nature, the processes of weathering is being carried out by various external agencies.

CHAPTER 6

GEOLOGICAL AGENTS

The surface features of the earth are the result of the action of various geological agents. It is well known that the landforms are the distinctive configurations of the land surface—mountains, hills, valleys, plains, and the like. In the evolution of landforms different exogenous as well as endogenous processes are mostly responsible. The configuration of the land surface are the outcome of the joint actions of some geological agents having both constructive as well as destructive effects on the existing surficial features.

The exogenous processes are those which derive their energy from external sources and ultimately from the sun. These processes are mainly caused by gelogical agents such as Blowing-wind, Running water, Glacier, Sea-Waves etc. Since these agents originate upon the earth's surface, they are known as *Epigene*—geological agents and their activities include processes like gradation, degradation, aggradation and weathering.

In a similar way, a process which originates within the earth's crust is termed endogenous. The geological agents, which are associated with these processes and have their origin underneath the surface of the earth are known as '*Hypogene*'—*Geological agents*. Earthquake, Volcanic eruptions as well as other earth-movements are the results of the hypogene-processes. Although they originate within, they do affect the earth's surface in a spectacular way. Mountains, plateaus and volcanoes are some of the striking features produced by hypogene agents.

Processes Associated With Epigene Geological Agents

Gradation. It is a three-fold process ; first the surface is decayed and eroded ; secondly the products of this decay and erosion are transported ; finally, they are deposited usually at lower levels. Thus 'Gradation' is the process by which the original irregularities of the earth's surface are removed and a levelled surface is created. All gradation processes are directed by gravity. Thus it is clear that processes of gradation are divisible into two major categories :

(*i*) Degradation.

(*ii*) Aggradation.

(*i*) **Degradation.** It constitutes those precesses by which material from a high relief feature is removed by exogeneous or external processes. It includes activities like weathering, mass-wasting and erosion including removal and transportation.

Factors Affecting Degradation

(*a*) Properties of the material on which the processes are operating,

(*b*) The energy available to the geomorphic agents, and

(*c*) The tools used by the agents.

(*it*) **Aggradation.** It is the process of elevation of low-lying tracts, through deposition of materials, by various epigene-geological agents like wind, river, glacier etc.

The epigene geological agents are grouped under two headings.

1. *On land.* Running water, ground water, wind, glacier etc.

2. *In oceans.* Waves, currents, tides etc.

The activities of the epigene-geological agents are divided into three major stages like, Erosion, Transportation and Deposition. These three activities contol the configuration of the land surface of the earth.

Hypogene processes. The two main endogenous processes are : volcanism and diastrophism. The geomorphic features produced by them provide the setting for the operation of the exogenous processes.

Diastrophism elevates or builds up portions of the earth's surface and are of two types : (*i*) Orogenic and (*ii*) Epeirogenic. While Orogeny refers to mountain building with deformation of the earth's crust ; epeirogeny refers to regional uplift without marked deformation.

The deformation of the crust is usually manifested in forms of folding, warping, faulting and with the occurrence of phenomena like continental drift, mountain building, volcanism, earthquakes isostasy etc.

Thus the hypogene and epigene-geological agents play their respective roles in shaping the surface features of the globe.

CHAPTER 7

VOLCANOES

Volcanoes are conical or dome-shaped structures built by the emission of lava and it contained gases from a restricted vent in the earth's surface. The volcanoes are having truncated tops representing the crater, that acts as the avenues for the magma to rise.

Volcanoes take many forms, and the activity that is associated with their eruption is highly varied. The activity of volcanoes differs in the amount and type of material ejected. The size, temperature and composition of the material ejected determine the shape of the volcano or the form of extrusion volcanoes also differ in the violence and the timing of successive eruptions.

Types of Volcanoes

1. A volcano is '*Active*' when it is erupting intermittently or continuously. A volcano which has not erupted for a long time is known as '*Dormant*', whereas an '*Extinct*' volcano is one, which has stopped eruption over a long time.

2. On the basis of mode of eruption as well as on the basis of nature of eruption, different types of volcanoes have been recognised.

Basing on the mode of eruption volcanoes are classified as :

(*i*) Central types, where the products escape through a single pipe (or vent).

(*ii*) Fissure types, where the ejection of lava takes place from a long fissure or a group of parallel or closed fissures.

Basing on the nature of eruption, volcanoes may be of two types as :

(*a*) **Explosive Type.** In which case the lava is of acidic (felsic) in nature and because of their high degree of viscosity, they produce explosive eruptions.

(*b*) **Quiet types.** In this case the lava is of basaltic composition (mafic lava), which is highly fluid and holds little gas, with the result that the eruptions are quiet and the lava can travel long distances to spread out in thin layers.

Besides the above, a number of other types of volcanoes have been identified according to their degree of explosive activity and nature of eruption. They are as follows :

(*i*) *Hawaiian type.* Silent effusion of lava without any explosive activity.

(*ii*) *Strombolian type.* Periodic eruption, with a little explosive activity.

(*iii*) *Vulcanian type.* Eruption takes place at longer intervals and the viscous lava quickly solidifies and gives rise to explosions of volcanic ash.

(*iv*) *Vesuvian type.* Highly explosive volcanic activity and eruption occurs after a long interval (measured in tens of years).

(*v*) *Plinian type.* The most violent type of vesuvian eruption is sometimes described as plinian. Here huge quantities of fragmental products are given out with little or no discharge of lava.

(*vi*) *Pelean type.* This is the most violent type of all the eruptions. They are characterised by eruption of '*nuées ardentes*'.

Volcanic topography. It includes both positive as well as negative relief features. The high or elevated relief features comprising of hills, mountains, cones, plateaus or upland plains are some of the examples of positive relief feature, while the low lying relief features like craters, calderas, tectonic depression etc. represent the negative relief features.

(*a*) **Positive-relief features.** These features are formed due to both quiet as well as explosive volcanic activity, and some of which are as follows :

(*i*) *Hornitos.* These are very small lava flows.

(*ii*) *Driblet cones.* The most acid lavas often give rise to quite small conelets and are known as driblet cones.

(*iii*) *Cinder cone.* These are volcanoes of central type of eruption, steep-sided with uniform slopes of 30° to 40°.

(*iv*) *Lava cone.* These are built up of lava flows, due to heaping of lava during quiet type of eruption. It is also known as lava or 'plug-dome'.

(*v*) *Composite cone.* These are made up alternatively of pyroclastic material and lava. Due to rude stratification, they are also known as 'Strato-volcanoes'.

(*vi*) *Shield-volcanoes.* These are made up of lava alone and due to quiet type of eruption, whereby piling up of flow after flow of fluid lava, a rounded dome like mass is produced.

(*vii*) *Spatter-cone.* Small cones formed on lava flows where breaks occur in the cooled surface of the flow allowing hot gases and lava to be blown out.

(*viii*) *Volcanic plateau.* These are formed because of fissure type of eruption.

(*b*) **Negetive-relief-features :**

(*i*) *Crater.* This is a depression located at the summit of the volcanic cone.

(*ii*) *Calderas.* Sometimes because of violent volcanic explosion the entire central portion of the volcano is destroyed and only a great central depression, named a 'caldera' remains. The calderas may also be formed due to erosion and enlargement of the crater.

(*iii*) *Lava-tunnels.* The more mobile lavas of basic composition, when erupted on the surface in the form of flows quickly consolidate and form a solid crust while the interior may still remain fluid. Under such conditions the enclosed fluid lava drain out through some weak spots lying at the periphery of the flow, forming what is known as '*lava tunnel*'.

(*iv*) *Cone-in-cone topography.* After an explosion destroys an existing crater, a new-built smaller cone with its own crater is built up. This is known as cone-in-cone topography.

Explosion-pits are also negetive relief features of volcanoes.

Volcanic products. A volcanic eruption comprises solid, liquid, and gaseous materials. Fragments of rocks ejected during an explosive eruption are called 'pyroclastic materials'. The pyroclastic materials of various size grades are known differently as follows :

(*i*) *Volcanic blocks (or bombs).* Diameter of the fragments is always above 32 milimetres.

(*ii*) *Cinders or lapilli.* Here the diameter is between 4 mm to 32 mm.

(*iii*) *Ash.* These particles range in size between 0˙25 mm to 4 mm.

(*iv*) *Fine-ash.* Minute particles of diameter less than 0˙25 mm constitute the fine ash.

(*v*) *Tuff.* Rocks made up of ash and fine ash are known as tuffs and the welded tuffs are known as 'Ignimbrite'.

(*vi*) *Agglomerates.* Pyroclastic rocks consisting mainly of fragments larger than 20 mm in diametre, are known as agglomerates.

Other Products of Volcanism

(*i*) *Porous or spongy.* Masses of solidified frothy lava is known as Pumice or Scoria. But scoria is having a little dark colour and coarse texture than that of pumice.

(*ii*) *Spilitic lava.* These are albitic (soda-rich) lava and it produces pillow structure.

(*iii*) *Gaseous product.* Steam forms the most important constituent of volcanic gases and contributes nearly 90% of the total content of volcanic gases. Gases like carbon-dioxide, nitrogen, sulphur-dioxide, hydrogen-sulphide, boric acid vapours, phosphorous, arsenic-vapour etc. forms a part of the volcanic gases.

(*iv*) *Nuée ardente.* An incandescent cloud of gas and volcanic ash, violently emitted during the eruption of pelean types of volcanoes.

Features Associated with the Decaying Phases of Volcanism

1. Fumaroles. These are fissures or vents through which volcanic gases are ejected.

Fumaroles emitting sulphurous vapour are called '*Salfataras*', and those which emit carbon-dioxide and boric-acid vapour are known as '*Mofettes*' and '*Saffoni*' respectively.

2. Hot springs. These are fissures through which hot-water escapes. The water usually gets heated with increased temperature below, may be magmatic or radioactive heat.

Calcareous deposits formed from hot springs are known as *Travertine* or *tufa* and similarly siliceous deposits are called *Siliceous sinters.*

3. Geyser. Hot springs ejecting boiling water and steam at regular intervals are geysers. Siliceous deposits formed around geysers are known as 'geyserite'.

Pseudo-volcanic features. Mud-volcanoes, meteor-craters are of non-volcanic origin and are examples of pseudo-volcanic features.

Cause of Volcanism

1. Release of high-pressures, which build-up within magma chambers below the ground surface.

2. Accumulation of radio-active heat produces magma ; of course other factors like frictional heat and the increase of heat with depth causes the formation of magma and their eruption on the earth's surface causes volcanism.

Distribution

1. They are concentrated in a narrow-belt called Circum-Pacific Ring of Fire, where the volcanoes are located on the high, young folded mountains.

2. Other volcanic areas include the scattered areas in the Pacific, particularly the Hawaiian Islands, a belt that includes Arabia, Madagascar and the rift valleys of Africa, the Mediterranean belt, the volcanoes of West-Indies and those of Iceland.

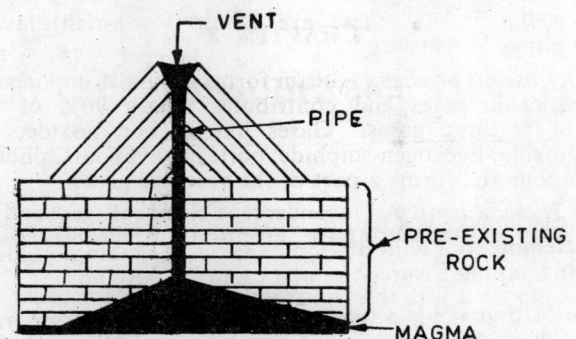

Fig 7·1. Cinder cone.

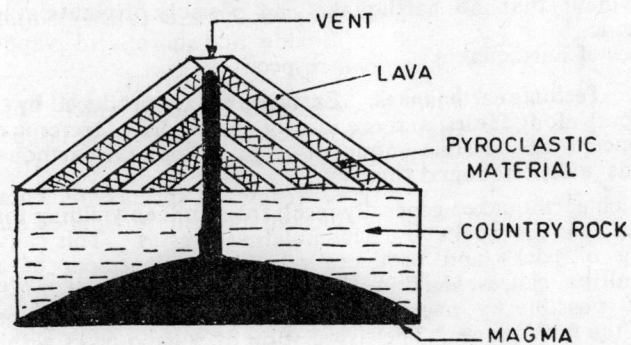

Fig. 7·2. Strato volcano.

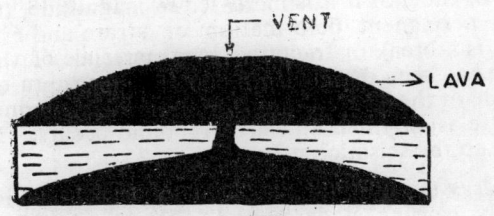

Fig, 7·3. Shield-volcano.

CHAPTER 8

EARTHQUAKES

An earthquake is a motion of the ground surface, ranging from a faint tremor to a wild motion capable of shaking buildings apart and causing gaping fissures to open up in the ground.

The earthquake is a form of energy of wave motion transmitted through the surface layer of the earth in widening circles from a point of sudden energy release—"*the earthquake focus*". It is, however, evident that no earthquake can possibly originate at a mere point alone.

I. Causes of Earthquakes

1. **Tectonic earthquakes.** Earthquakes are produced by sudden movements along faults, and are mostly, therefore of tectonic origin. The concept of possible mode of origin of tectonic earthquakes is known as 'elastic-rebound theory'.

Such earthquakes generally result from sudden yielding to strain produced on the rocks by accumulating stresses. This causes the breaking of rocks and produces relative displacement of rocks. Such faulting causes shaking because displacement of rocks can only be possible by overcoming frictional resistance against the walls of the fault-plane. The association of earthquakes with fault-lines is an established fact.

Elastic-rebound theory. According to **Prof. H.F. Reid**, materials of the earth, being elastic, can withstand a certain amount of stress without deforming permanently, but if the stress is continued for a long period of time, or if it is increased in magnitude, the rocks will first take a permanent deformation or strain and eventually rupture. A fault is a break or fracture in the materials of the earth along which there has been displacement. When the rupture occurs, rocks on either side of the fault tend to return to their original shape because of their elasticity and an elastic rebound occurs. It is this rebound that sets up the seismic waves.

Thus the energy stored in the system through decades, is released instantaneously causing underground dislocation of rocks and setting up vibrations which are feeble or strong, as the case may be.

2. Volcanic earthquakes. Usually, earthquakes associated with volcanoes are more localized both in extent of damage and in intensity of the waves produced in comparison to those which are associated with faulting motions. A shock may be produced by any of the following mechanisms :

(*a*) explosion of the volcano upon the release and expansion of gases and lavas,

(*b*) faulting within the volcono resulting from pressures in the chamber of molten rock, and

(*c*) collapse of the centre of the volcano into the space formed by the extrusion of gases and molten matter.

Besides the above, sometimes local reasons may cause feeble earthquakes, like failure of dams under the pressure of the impounding water etc.

II. Terminologies Associated with Earthquakes

(*a*) **Focus.** It is the exact spot underneath the earth's surface at which an earthquake originates. It is also known as Hypocentre.

(*b*) **Epicentre.** It is the point on the surface of the Earth, above the focus of an earthquake.

(*c*) **Isoseismal line or isoseists.** It is a line joining all points at which the intensity of the earthquake is the same. It is, in fact, an isodiastrophic line of equal damage.

(*d*) **Homoseismals or coseismals or homoseists.** These are lines joining the places where the shock arrives at the same time.

III. Types of Earthquakes

Natural earthquakes are of three types ; according to the depth of their origin. They are as follows :

(*a*) **Shallow-focus earthquakes.** In this case the seismic shocks originate at a depth of about 30 miles or less, below the earth's surface.

(*b*) **Intermediate-focus earthquakes.** In this case the shock-waves originate at a depth between 30 to 150 miles.

(*c*) **Deep-focus earthquake.** Here the point of origin of the shock is at a depth between 150 to 450 miles.

According to the origin of the earthquakes, they are also of three types like—tectonic, volcanic and submarine earthquakes. The submarine shocks often generate very large waves on the surface of the seas and destroy the coastal tracts. These submarine earthquakes are known as '*Tsunamis*'.

Three main types of wave-motion are generated by an earthquake :

(*a*) **Primary or 'P' waves.** These are longitudinal waves similar to sound waves and travel in solid, liquid and gaseous media. They have short wavelength and high-frequency.

(*b*) **Secondary or 'S' waves.** These are transverse waves, also known as sheer waves, travel only in solid media. In comparison to primary waves, they are slow in motion. They travel at varying velocities through the solid parts, proportional to the density of the materials. They are also having short wavelength and high frequency.

(*c*) **'L' waves.** These are transverse vibrations and are confined to the outer skin of the crust. They are also known as surface waves or Rayleigh (R) waves. They have low velocity, low frequency and long-wavelength. These are responsible for most of the destructive force of the earthquake.

IV. Scale of Intensity

Various scales have been proposed to estimate the intensity of earthquake from the amount of damage caused. These scales are :

(*i*) Rosi-Forrel scale,

(*ii*) Mercalli-scale, and

(*iii*) Richter scale of earthquake magnitude etc.

In the Rosi Forrel scale, the intensity has been classified into—severe, catastrophic and disastrous.

The Mercalli intensity scale has devised twelve numbers with the increase of intensity. In this case—number 1 detected only by seismographs.

Gradually the number increases when the earthquake intensity becomes feeble, slight, moderate, strong etc. At number '8' it is "destructive". Similarly it becomes "disastrous" at number '10', and at number '12', the effect is totally catastrophic, where there is total destruction and objects thrown into air.

In the Richter-scale, the scale number ranges from '0' to '9'. Here it is particularly important to notice that a magnitude—'8' earthquake is 10 times larger than a magnitude—7 earthquake, 100 times larger than a magnitude—6-earthquake, and 1000 times larger than a magnitude—5-earthquake.

V. Recording of Earthquakes

The instrument used for recording of seismic shocks is known as '*Seismograph*', and the records of seismic shocks prepared and presented by seismographs are known as '*seismograms*'.

VI. Distribution of Earthquakes

The destructive earthquakes are concentrated in a ring surrounding the Pacific Ocean. This ring coincides with the Circum-Pacific Ring of Fire.

The second chain is termed as East-Indian, which extends over Indonesia, Andaman-Nicobar, Islands and Burma.

The third belt extends over Himalayas, Kun-Lun, Tien Shan and Altai Range up to the Lake Baikal.

Another belt extends from the Pamir Knot to Afghanistan, Iran, Turkey, Greece, Rumania, Atlas Mountains, Gibralatar and the Azores Islands.

A belt also extends from the Gulf of Aden, between Seychelles and Maladive Islands, turns to the West-South of Africa and goes up to the Falkland Islands.

Another belt also runs along the Great Rift Valley of East Africa.

It is noticed that the present earthquake regions are associated with the younger fold-mountain regions, and the present earthquake activity is a phase of the end of the Alpine-Orogeny.

CHAPTER 9

GEOSYNCLINES AND MOUNTAINS

1. Geosynclines. The 'Geosynclines' are major structural and sedimentational units of the crust of the earth. They are elongated trough-like depressions submerged beneath the sea-water. They are considered to be the future sites of mountain building activity and of fold-mountains.

These basins become filled with very great thickness of sediments and alongwith the accumulation of pile of sediments, there occurs progressive subsidence of the basin floor.

2. Types of geosyncline. There are seven types of geosynclines as follows :

(*i*) **Ortho-geosyncline.** These are elongated basins which become filled with very great thickness of sediments, which is subsequently deformed to form a fold-mountain chain.

(*ii*) **Eugeosyncline.** In these geosynclines, the pile of sediments are found with an abundance of volcanic rocks ; they are formed at some distance from the shield areas, *i.e.*, Kratons.

(*iii*) **Miogeosyncline.** These are formed adjacent to the Kraton, where there is a thinner development of sediments which lack volcanic rocks.

(*iv*) **Taphrogeosyncline.** This is an elongated depression, formed because of faulting. These are also known as 'graben' or 'rift-valley'.

(*v*) **Parageosyncline.** This is the geosyncline which lies within the Kraton.

(*vi*) **Zeugo-geosyncline.** These are parageosyncline with marginal uplifts.

(*vii*) **Auto-geosyncline.** It is a para-geosyncline without marginal uplifts.

Most orogenic-belts arise on the sites of geosynclines and the resulting mountains therefore consist of sediments and volcanic rocks deformed and metamorphosed to a greater or lesser extent according to their position and depth in the orogenic belt.

The fact that the sedimentary units are thicker along the mountain belts than outside indicates that most sediments now exposed in mountain chain were deposited in geosynclines. True geosynclinal accumulations presently lie off the eastern and gulf-coasts of America under the continental shelves and continental slopes.

1. **Mountains.** Mountains are isolated or interlinked masses of land elevated appreciably above the average altitude of their surroundings and are characterised by the presence of pointed or ridge-like tops known as 'peaks'.

2. **Terminologies associated with mountains :**

(*i*) The smallest unit of an elelvated land-mass is known as a '*hillock or mound*'.

(*ii*) Larger elevated land-masses are know as '*hills* or *mountains*'. But mountains are usually 1,000 metres high or above, and those which are having less heights are the 'hills'.

(*iii*) A series of inter-connected mountains constitute a '*range*'.

(*iv*) A number of genetically related ranges, together form a '*system*'.

(*v*) A few systems combine to form a '*Chain*'.

(*vi*) Several inter-related chains form a '*Cordillera*'.

3. **Types of mountains.** On the basis of their mode of origin, mountains can be grouped into three categories as follows :

(*a*) Mountains of accumulation.

(*b*) Relict or residual mountains.

(*c*) Tectonic mountains.

Mountains of accumulation are due to accumulation of volcanic materials like lava and other pyroclastic materials, into heaps or cones, which produce volcanic mountains. Sand-dunes formed due to accumulation of piles of sands belong to this category.

Relict mountains are because of differential erosion, which is noticeable in regions which are composed of rocks of various strength and durability.

Tectonic mountains are formed because of the roles played by the Diastrophic forces, and are of three types as :

(*i*) **Fold-mountains.** The bending of rock strata due to compressional forces acting tangentially or horizontally towards a

common point or plane from opposite directions is known as folding;
which results in fold-mountains.

(*ii*) **Fault-mountains.** These are also known as Block-mountains,
which are because of faulting. Horst or a block mountain is an
uplifted landmass located between two adjacent faults. A graben
or rift valley is the block lowered between two adjacent faults.

(*iii*) **Dome-mountains.** Igneous instrusions, sometimes, make
room for themselves by lifting up the overlying layers of the country
rocks. Thus a dome-shaped structure is formed, which are sometimes large enough to be described as dome-mountains *e.g.*,
laccolith.

4. **Causes of mountain building.** So many hypothesis have been
put-forwarded to explain the origin of mountains. Some of the
important hypothesis may be described as follows :

(*a*) Geosynclinal hypothesis

(*b*) Theory of isostasy.

(*c*) Contraction hypothesis.

(*d*) Convection-current hypothesis.

(*e*) Continental drift hypothesis.

(*f*) Theory of plate-tectonics.

(*a*) **Geosynclinal hypothesis.** Some hold that, under the load of
sediments the floor of the subsiding geosynclinal basin is likely to
be broken and thus the sediments would meet the internal heat and
expands in volume. As a result of which the upper sendimentary
layer would be uplifted. But this process does not account for the
compressional forces.

Others believe that the geosynclinal basin on receiving the
lodes will sink down and in this process, the two sides of the
shallow trough will be brought nearer. This will generate compressional forces and may account for the formation of fold mountains.

(*b*) **Theory of isostasy.** Isostatic adjustment is believed to play
an important role in mountain building, but this process only
accounts for vertical upliftment and not the compressional forces.

(*c*) **Contraction hypothesis.** It states that earth through the
radiation of heat is cooling and as a result the earth will shrink and
wrinkles will be developed on its surface, just as a mango if left
exposed to heat will shrink and develop wrinkles upon its surface.
But it does not explain the occurrence of fold-mountains along a
few-belts only, besides the fact that radioactive substances produce
heat on disintegration has not been taken into account.

(*d*) **Convection-current hypothesis.** It is postulated by Holmes
and Griggs. It is believed that the earth has a thin solid-crust and

a metallic core, while the intervening layer is made up of molten silicates. Since silicates are bad conductors of heat particularly in the molten state, it transmits heat by convectional process.

In this process, where currents are flowing horizontally under the surface of the crust, they exert a powerful drag on the crust and throw it into tension where they diverge and into compression, where they converge. Thus we should expect orogenic belts to be formed where two approaching currents turned down.

(e) **Continental drift hypothesis.** According to this hypothesis, the sialic blocks during their course of movement through the sima are affected by the resistance of sima. This resistance crumple and throws the frontal moving parts of the continent. According to Wagener, the Rockies and Andies are formed in this way, by the Westward drifting of the continents. The Himalayas and the Alpine mountain belts are thought to have been formed by the equatorial movement of India and Africa.

(f) **Theory of plate-tectonics.** It explains that the top-crust of the earth is a mosaic of several rigid segments called plates, which include not only the solid upper crust but also part of the denser mantle below. They float on the plastic upper mantle known as 'asthenosphere'.

Plates may diverge, converge or move in parallels. Plates are said to diverge when two adjacent plates move apart and hot magma comes up through the crack and solidifies, accounting for the formation of mid-oceanic ridges. Plates are said to converge when they come together and collide. When two continental plates converge, the lateral pressure exerted by them crumples and compresses them into folds. When this crumpling and folding continues high mountains form.

Accordingly, the Himalayas were formed because of the northward movement of the Indian-plate against the Chinese-plate.

All the aforesaid hypothesis explain the origin of mountains in some or other ways.

CHAPTER 10

ISOSTASY

The word 'Isostasy' is derived from a Greek word meaning 'in equipoise' or 'in balance'. The theory of 'Isostasy' postulates a system for the distribution of materials in the earth's crust which conforms to and explain the observed gravity values. This theory was developed from gravity surveys in the mountains of India (1850). The term was first used by the American Geologist—Dutton in 1889.

This doctrine states that wherever equilibrium exists in the earth's surface, equal mass must underlie equal surface areas ; in other words, a great continental mass must be formed of lighter materials than that supposed to constitute the ocean floors and further, in order to compensate for its greater height these lighter continental material must extend downward to some distance under the continent and below the ocean floor level in order that unit areas beneath oceans and continents may remain in stable equilibrium.

According to Dutton, the elevated masses are characterised by rocks of low density and the depressed basins are characterised by rocks of higher-density. Accordingly, a level is thought to exist where the pressure due to elevated masses and the depressed areas would be equal. This is called 'Isopiestic Level' or the 'Level of Uniform Pressure'.

Any loading due to sedimentation, deposition or intrusion of igneous material etc. or unloading due to denudation or melting of ice will, therefore, disturb the balance and compensation in the form of depression or elevation will follow to restore the state of hydro-static balance. The state of isostasy can be maintained only if there is a continuous compensation at depth.

Through erosion, material is being removed from the tops of the mountains, which therefore are becoming lighter. Materials should therefore move into the roots of the mountains at depth through the interior of the earth. This movement is termed compensation.

The zone between the isopiestic level and the surface of the earth is called the zone of compensation or lithosphere. The zone below the isopiestic level is called the asthenosphere.

In analyzing the concept of equal pressure or weight, it follows that the blocks in balance have equal masses. If granites and basalts are considered, in the light of equal mass concept, the granites being 10% lighter than the basalts, must have higher volume which is represented by the highlands composed of sialic matter ; on the contrary the volume of the denser matter, *i.e.*, basalt or sima has ought to be less and that is why they are represented by depressions or ocean basins.

In this regard, three theories have been (put forwarded) as follows :

1. Airy theory. According to this theory, there is a change in the density of rocks at depth in the earth's interior and that the upder lighter material floats on the dense layer. The depth of this change varies from place to place. Airy preferred to assume that the different crustal blocks are of equal density and unequal thickness. According to him, the blocks constituting the mountains are thicker than those on which the plains lie and as a result they stand higher up just in the same way in which a huge block of ice projects much above the smaller ones when allowed to float freely in water. The thicker block at the same time sinks deeper down in the sub-stratum and thus constitutes the root of the corresponding mountains.

According to Airy therefore mountains are supported by their roots which penetrate deep down into the denser sub-stratum and keep the mountain floating due to its buoyancy.

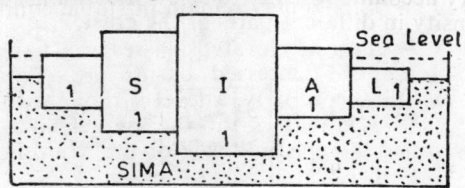

Fig. 10˙1. All blocks have the same density.

This has greater support from recent geophysical data. For example, Mt. Everest in the Himalayas rises to a height of about 9 kms, right beneath it the crust is about 80 kms thick ; whereas the crust is only 32 kms thick in the Deccan Plateau and only 12 to 20 km or less near sea-coasts of India.

2. Pratt-theory. According to this theory, the additional mass of a great mountain is compensated by a corresponding deficiency in the density of the rocks constituting the crustal block on which the mountain stands.

Here, it has also been assumed that the boundary between the upper light material and the lower dense rocks is at a uniform depth, called the depth of compensation. Besides, it is also presumed that there are variations in the density of the lighter layers which are related to the elevation of the surface. The weight of columns of rock extending from the surface to the depth of compensation in different parts of the earth is thus the same.

Observations made with the help of seismic studies indicate that more dense material underlies ocean basins than continents, which support Pratt's assumptions.

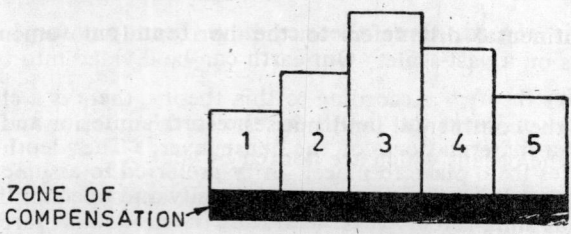

ZONE OF
COMPENSATION

Fig. 10·2. Higher blocks have a lower density than the lower blocks.

3. Heiskanen's theory. He combines the assumptions of both Airy and Pratt. It is observed that rocks at sea-level are more dense on the average than those at higher elevations. He assumes this change continues downwards, tending to make deeper rocks more dense than the shallower ones. In addition, different sections are thought to have different densities and different lengths.

This theory accounts for the roots of mountains and for the variations in density in different parts of the crust.

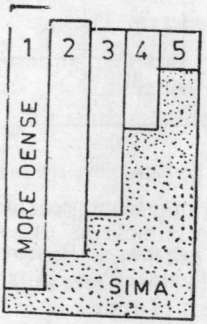

MORE DENSE

SIMA

Fig. 10·3. Density varies between columns and within each. Higher densities are represented by the shorter columns and towards the bottom of each column.

CHAPTER 11

CONTINENTAL-DRIFT

Continental drift refers to the horizontal movement of the continents on a vast-scale. Our earth can be divided into two major units :

    (*i*) the continental land masses, essentially composed of sial.

    (*ii*) the oceans which are floored by sima.

It is assumed that the continental masses are floating above the viscous sima-layer. It is well-known that there is a remarkable accumulation of land masses in the northern-hemisphere of the earth, while there is appreciable concentration of water bodies in southern-hemisphere. It has been further noticed that some of the continental masses like Africa, South America etc. exhibit southerly tapering edges. The major problem of geology is to what extent these arrangements have been stable during the geological past.

The evidences in favour of the 'Continental-drift' hypothesis were collected from the physiographic, geological and meteorological observations. The following evidences strengthen the idea of 'Continental-drift' :

1. Geological Evidences. These are as follows :

    (*i*) Continental boundaries on either side of the Atlantic-ocean are matching with each other.

    (*ii*) Similarity in fossil contents *i.e.*, faunal and floral similarity.

    (*iii*) Lithological similarity.

    (*iv*) Stratigraphic similarity, with relation to age.

2. Geo-physical evidences. Palaeo-magnetic studies and geodetic survey gives evidences in support of the phenomenon of Continental drift.

3. Tectonic evidences. These are as follows ʲ

    (*i*) Geological structures.

    (*ii*) Development of mid-oceanic ridges.

(*iii*) Development of fold-mountains.

(*iv*) Development of island-arcs.

4. Meteorological evidences. Information obtained from palaeo-climatic and palaeo-geographic analyses, indicate that drifting of continents had taken place during the geological past.

The idea of continental drift was develeped by F.B. Taylor in America, and Alfred Wegener in Germany. Their theories are as follows :

Taylor's hypothesis. According to him long back there were two great landmasses in the southern and northern hemisphere known as Gondwana land and Laurasia, respectively. He supposed that in due course of time those land masses started spreading towards the equator, more or less radially from the polar-regions.

He assumed that the sudden increase of tidal action of the moon in late cretaceous time, increased the rate of rotation of the earth which dragged the continents away from the poles. Accordingly he cited certain evidences in support of his theories, that (*i*) East-coast of South America is similar in shape to the West coast of Africa and they fit like the parts of a 'Zig-saw'. (*ii*) Location of most of the tertiary mountains are found more or less towards the equator side of the continental masses.

But his assumptions fail to justify the exact-happenings ; for example tidal forces never increase the earth's rotation, but the assumption is just the reverse of it. Besides, the equatorial movement of continents as assumed by Taylor fail to explain the drifting of South America from Africa.

Wegener's hypothesis. According to him all the sialic layer was concentrated in a large continent called the 'Pangea' before the Silurian period. This block floated in a universal basaltic layer. Pangea was surrounded by the world-ocean called 'the Panthalassa'.

In the late Palaeozoic period, probably during permian or Mesozoic era, the Pangea broke into pieces under the influence of the tidal force, and the force generated by the movements of the earth's axis of rotation and revolution.

The southern parts of the Pangea broke apart during Mesozoic and the northern in the Tertiary periods. The Continental drift was caused by the differentials gravitational forces which acted upon the protruding-block of sial. One force caused the drift towards the equator and the other towards the West.

The African-block (the Gondwanaland) and the Eurasian-block (the Laurasia) moved towards the equator. When the drift towards the equator was taking place, the Americas drifted towards the West. Thus the Atlantic Ocean was created between North and

South America in the West and Europe and Africa in the East. Australia was left behind in the beginning. Later it swung to the east. Only recently, in terms of geological time [periods Antarctica separated from South America.

Before drifting, North and South America formed one unit. They rotated about a point in North America. Then they were drawn apart. This produced the narrow land of central America and the scattered fragments of the West Indies. Labrador and New Foundland separated from Europe during Quaternary. They swung South-West and Greenland was left behind as a separate-block. At the same time the Indian part of the Gondwanaland moved north against the mass of the Asian main continent. It separated from Africa. Madagascar was left behind. By the compression of the Indian part against the Angara shield (*i.e.*, Asian main continent) the mountain chains of the Himalayas were created.

Wegener and his followers, in favour of this theory, have put forwarded the following evidences :

(*a*) Similarities of the coast-lines once thought to be adjacent, especially South America and Africa.

(*b*) The similarity of pre-Cambrian (before 500 million years ago) rocks of Central Africa, Madagascar, Southern India, Brazil and Australia.

(*c*) The continuity of tectonic trends of the blocks of these countries across their present boundaries.

(*d*) There are wide-spread occurrences of carboniferous coal deposits implying a tropical humid-climate. The coal bearing formations are now found in South America, South Africa, Madagascar, India, Australia and Antarctica.

(*e*) Unmistakable evidence of wide-spread glaciation towards the end of the Palaeozoic era is found at the southern extremity of South America, the southern-half of Africa, in the peninsular India, extending to the Himalayan regions and in Australia.

Although the theory of continental drift has been widely discussed and accepted, serious doubts have been raised about the period during which the force causing the drift had operated and also about the direction and the amount of force. However the recent theories of 'Plate-tectonics' and information obtained from palaeo-magnetic studies, lends support to the theory of Continental drift.

Palaeo-magnetism. According to the studies on palaeo-magnetism, it is possible to find out the direction and dip of the earth's magnetic field during different geologic periods. It has been observed that the pole-positions of the present globe are different during the geologic past and by joining these poles, a curve is obtained, which is known as 'polar-wandering curve'. It is seen that the polar wandering curves drawn for different continents are not parallel or sympathetic, which confirms 'Continental-drift'.

Palaeomagnetic works on the Deccan plateau-basalt shows that the average rate of movement with refrence to India is about 7 cm/year.

Plate tectonics. The concept of the movement of lithospheric plates and the sea-floor spreading, along with their supporting evidences prove that there was drifting of continents.

A general concept of continental drift may be obtained from the following stages :

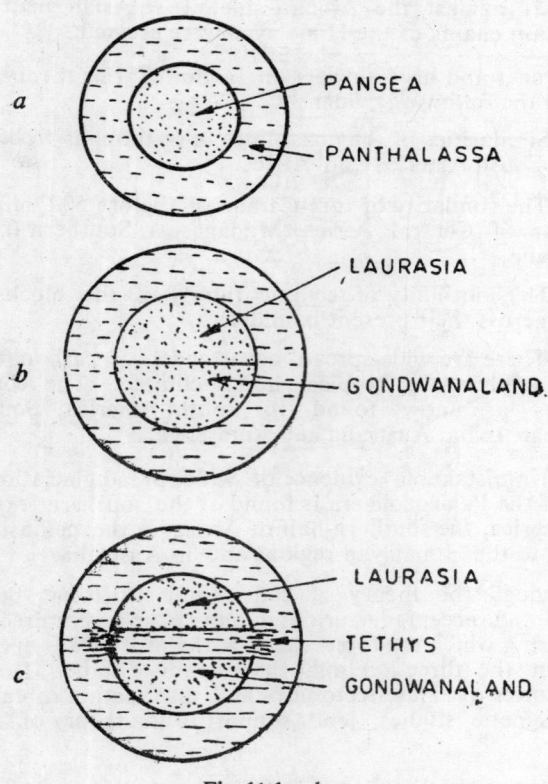

Fig. 11·1 a, b, c.

during the positive part of the cycle. At the node, a curve is
.... It is seen that
.... chromosomes are not

.... shows
.... about

.... boundaries

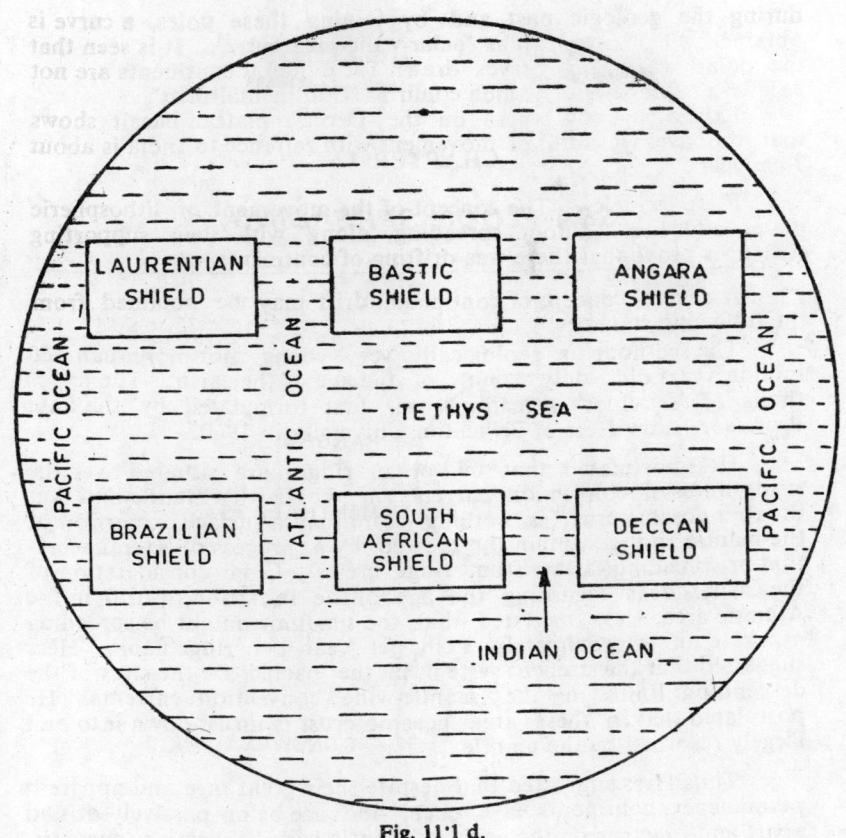

Fig. 11·1 d.

CHAPTER 12

SEA-FLOOR SPREADING

The sea-floor is geologically very young, not more than 160 million years old ; only 1/30th of the age of the earth. The hypothesis of 'sea-floor spreading' was first formulated by the Late Professor Harry Hess of Princeton University in 1960.

Hess postulated that mid-ocean ridges are situated over the rising limbs of convection currents in the Earth's mantle and that the thin oceanic crust is nothing more than a surface expression of the mantle and continuously created by a process of lateral accretion or spreading away from ridge crests. From consideration of the earlier ideas regarding the age of the initiation of drift in the Atlantic area, Hess suggested that the sea-floor might be spreading at a rate of approximately 1 cm per year per ridge flank. Hess suggested that the trench system in the pacific are the sites of the descending limbs of the mantle-wide convection currents. He postulated that in these areas oceanic crust is thrust down into and largely resorbed by the mantle.

Thus Hess suggested that despite their great age and apparent permanency, continents have been, and are being passively drifted apart and together on the backs of mantle-wide convection currents. In contrast, the ocean-floors are young and ephemeral features of the Earth's surface, constantly being regenerated at ridge crests and destroyed in the trench systems.

Supporting evidences. (*i*) The occurence of earth-quakes along the crest of the mid-oceanic ridge system, the dearth of sediments at ridge crests and the active volcanic islands associated with the crest of the Mid-Atlantic ridge are all readily explained by Hess' model.

(*ii*) Moreover the ocean basins as a whole contain a remarkable thin veneer of sediments and small number of sea-mounts, if recent rates of accumulation and formation are extrapolated over the whole of geologic time.

(*iii*) In 1960, Hess calculated that South America and Africa had both moved 2500 km from the mid-atlantic ridge during an interval which he thought was 250 million years. This gives the rate of separation as 10 mm/year. **Rates like 10 mm/year are appreciable in human terms.**

(*iv*) Hess was also able to state that no material greater than about 100 million years in age had ever been recorded from the deep ocean floor or truely oceanic islands.

(*v*) Hess argued that the anomalous high values of heat reflect the emplacement of hot mantle-derived material in the vicinity of ridge crest.

(*vi*) Marine-geophysical studies have also revealed that ridge-crests are characterised sometimes by anomalously low seismic wave velocities in the upper-mantle. This was attributed to thermal expansion and micro-fracturing associated with the upwelling mantle, both effects producing a reduction in the seismic wave velocities and density of the mantle-material.

(*vii*) Palaeomagnetic studies have also revealed that new sea-floor forms at and spreads laterally away from ridge crest.

These evidences have been accumulated to show that the crust is spreading apart along the rift. As this sea-floor spreading occurs basaltic lava rises from beneath the rift, solidifying and forming new oceanic crust.

CHAPTER 13

PLATE-TECTONICS

It is known that the crust of the earth consists essentially of about 35 km thick layer of solid rock matter, which varies in thickness from about 5 km in the oceanic areas to even such thickness as 70 to 80 kms in the mountainous regions of the Alps and the Himalayas. That the crust is not fully rigid, but has been repeatedly deformed in the geologic past and is subjected to movements even now is proved by earthquakes frequent on the ocean-floors and rarer in continents, volcanism, folding and faulting of large expenses of rock strata and recent elevation and depression of coastal areas.

There are now tangible evidences chiefly from palaeomagnetic studies that the crust of the earth (oceanic and continental) together with the upper portion of the mantle, which overlies the asthenosphere (the low-velocity zone approximately at 100 to 150 kms depth) constitute the lithosphere, which is disjointed into plates or blocks extensively by faults or thrusts. Thus the lithosphere is made up of lithospheric-plates.

Plate-tectonics provides a modern view of the rock-cycle. It involves a world-wide net-work of moving plates of lithosphere.

Important Features of Plate-Tectonics

(*i*) It is assumed that the earth is composed of 20 lithospheric plates.

(*ii*) They have the thickness from 0 to 10 kms at the ridges to 100-150 kms elsewhere.

(*iii*) The plates may contain continental as well as oceanic surfaces.

(*iv*) These plates are continuously in motion both with respect to each other and to the Earth's axis of rotation.

(*v*) Virtually all seismicity, volcanicity and tectonic activity is localized around plate margins and is associated with differential motion between adjacent plates.

(*a*) **Plate boundary**. It is the surface trace of the zone of motion between two plates.

(*b*) **Plate-margin**. The marginal part of a particular plate.

Two plate margins meet at a common plate boundary.

(*vi*) These plates are small and large, separated by faults and thrusts, lying mostly across ridges or parallel to the continental borders (trenches).

(*vii*) They move with velocities ranging from 1 to 6 cm per year.

(*viii*) Where two plates diverge, we find extensional features, typically the oceanic ridges, symmetrical about the vertical axis.

(*ix*) Where two plates slide past each other, we find transcurrent faults, *i.e.*, the large strike-slip faults joining segments of ocean ridges or arcs.

(*x*) Where two plates converge, and one is thrust beneath the other we find the island arcs, the huge assymmetric features that are the sites of greatest earthquakes, explosive volcanism, great topographic relief and many other distinctive features.

Type of plate-margin. There are three types of plate margin :

(*a*) Constructive, (*b*) Destructive, and (*c*) Conservative.

(*a*) **Constructive**. In this case new crust is created by the upwelling of materials from the mantle. The lithospheric plates. diverge at the crest of the mid-oceanic ridges where new surface is created. Thus a ridge represents a zone along which two plates are in motion away from each other, yet they do not separate because new material is continuously added to the rear of each. Boundaries at which the net effect of motion is to generate surface area are termed as 'Sources'.

In case of constructive plate boundaries, the greatest principal stress is vertical and the plate boundary will consist of a set of normal faults dipping about 60° from the horizontal.

(*b*) **Destructive**. It represents the zone of convergent plate boundaries, along which two lithospheric plates are coming together and one plate is forced to plunge down into the mantle. In this case the plate boundary will be a reverse fault dipping at angles of 30° from the horizontal. This overriding of one plate on another gives rise to trenches and island arcs. Plate boundaries at which the net effect of the motion is to destroy surface area are called 'Sinks'.

(*c*) **Conservative margins**'. When the lithospheric plates can slide past one another and that the plates neither gain nor lose surface areas, there results a transcurrent or transform fault, which marks the conservative plate boundaries.

Although plates may comprise either continental or oceanic crust or both, it seems that only those parts of plates which are capped by oceanic crust can participate in the main processes of plate-growth and destruction.

Euler's theorem. It is a geometrical theorem which shows that every displacement of a plate from one position to another on the surface of a sphere can be regarded as a simple rotation of the plate about a suitably chosen axis passing through the centre of the sphere. All points on the plate travel along small circular paths about the chosen axis in passing from their initial to final position. It follows that any plate boundary which is conservative must be parallel to a small circle, the axis of which is the axis of rotation for the relative motion of the plate on either side. Conversely, any plate boundaries which are not parallel to such small circles must be either constructive or destructive.

To understand the mechanisms of the movement of the plates, it is necessary to know the physical characters of the lithosphere :

(*a*) **Thermal property.** The lithosphere strongly modifies the stress and temperature fields as it transmits them from the asthenosphere to the earth's surface. As the hot newly created, lithospheric plate moves away from the accreting plate boundary, it progressively cools according to an exponential law through flow of heat at its surface.

(*b*) **Elasticity.** The lithospheric plate can be considered to be a thin elastic sheet which floats over a fluid substratum and bends under super-crustal load.

(*c*) **Mechanical properties.** The newly created oceanic lithoshpere near the mid-oceanic ridges is hot and very thin and should be much weaker than the normal ocean bearing or continent bearing lithospheric plate. However at equality of thickness, continent bearing plates are easier to deform than ocean-bearing plates.

(*d*) **Lithosphere as a stress-guide.** The zone of deep and intermediate earthquakes often called Benioff zones, correspond to stresses occurring within a lithospheric plate which sinks into the asthenosphere and not to faulting between the asthenosphere and the lithospheric plates.

The tectonically active zones are now referrable to boundaries of six major and about twice as many minor, more or less rigid inter-plate.

(*e*) **Kinematics of relative movements.** According to Wilson the lines of creation of surface produce surface symmetrically, while lines of destruction of surface destroy surface asymmetrically. These lines could end abruptly against what he called a transform fault, along which the movement was pure strike-slip.

Causes of Plate Motion

1. Oceanic crust formation. As the crustal formation at the mid-oceanic ridges is a continuous process, it begins to spread at a rate of 1—6 cm/year and this may cause the motion of the plates.

2. Rates of motion. Since spreading occurs at ridges, at rates ranging from 1—6 cm/year but crust is consumed at a rate of 5 to 15 cm/year at oceanic trenches, the plates are to move.

3. Oceanic-topography. The mechanism must be consistent with the development of topographic ridges at centres of spreading ; ridges rises 2 to 4 kms above the level of the ocean floor and near the axes slope away more or less symmetrically from the crest.

4. Gravity. Ridges are close to isostatic equilibrium but sinks are characterised by topographic trenches which shows the largest negative gravity anomalies. The gravitational difference may cause plate motion.

5. Thermal. With increasing distance from ridge crest the scatter of heat flow values diminish and the mean heat-flow falls until it reaches average level for the oceans. Oceanic trenches have abnormally low heat flow but a short distance away in the adjacent island-arcs, the flow is high. The difference between these heat flow values seem to be responsible for plate motion.

6. Convection-current condition in the mantle zone seems to be responsible for plate motion, as the diverging current drags the lithospheric plates along the direction of their flow.

7. Strength of the lithosphere. Even though lithospheric plates appear to be able to move great distances without undergoing significant internal deformation, in some instances the plates are 20 times as long as they are thick. With such a length to thickness ratio neither compressional nor tensional stresses could be transmitted from one end of the plate to the other, unless the frictional resistance beneath the plates are very small. As the plates are above the viscous melt, it suffers no friction and can move.

Importance of plate-tectonics. The theory of plate tectonics is useful in explaining the phenomena like :

(*i*) Continental-drift.

(*ii*) Mountain building (where two continental plates collide with each other).

(*iii*) Island-arcs (where an oceanic plate undergoes subduction beneath a continental plate).

(*iv*) Oceanic trenches.

(*v*) Ocean-floor spreading.

(*vi*) Mid-oceanic ridges (where plates diverge).

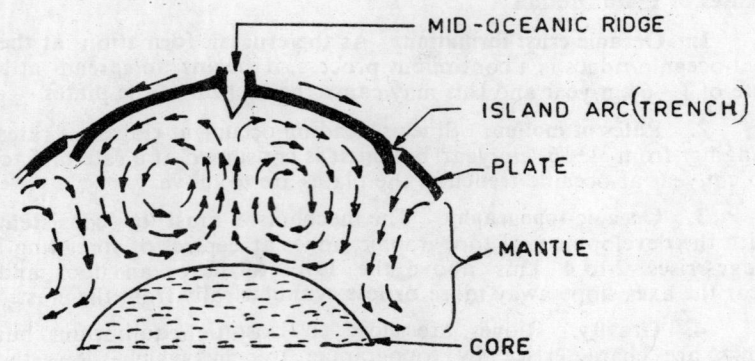

Fig. 13·1.   Convection system in the mantle causing motion
of the overlying lithospheric plates.

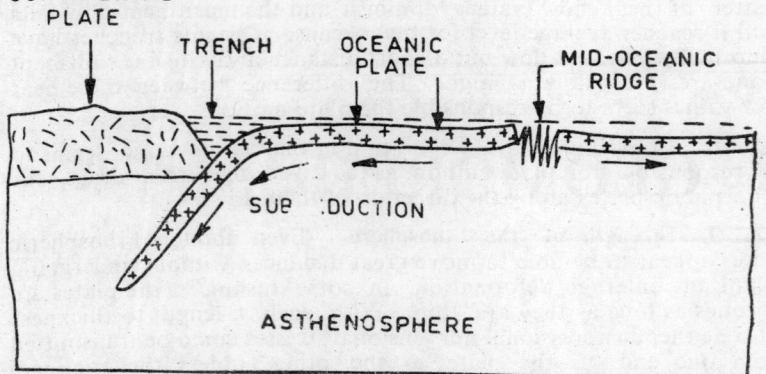

Fig. 13·2.   Illustration of plate motion. Oceanic plate glides over
the soft asthenosphere from mid-oceanic ridge to the
zones of subduction.

PART II
GEOMORPHOLOGY

CHAPTER 14

BASIC-CONCEPTS OF GEOMORPHOLOGY AND TYPICAL-LANDFORMS

Geomorphology is the systematic study of land-forms and the interpretation of them as records of past-history. Certain agents function to bring about gradation and result changes in land-forms. These may be called geomorphic agents, such as river, wind, glacier and sea-waves. These geomorphic agents basically serves three functions ;

(*i*) Erosion,

(*ii*) Transportation, and

(*iii*) Deposition.

(*i*) **Erosion**. The breaking-down or disintegration of rocks due to the physical forces associated with the natural agencies, followed by removal of the dislodged rock-fragments and particles, and the sum total of the process is defined as erosion.

(*ii*) **Transportation**. It is the process by which the weathered materials are removed from the site of their formation by various geomorphic agents.

(*iii*) **Deposition**. It is the process, whereby accumulation of the transported materials, which results from the loss of the transporting capacity of the geomorphic, agents. This happens when the velocity is reduced or the material being transported exceeds the capacity for transporting. It may also happen because of chemical precipitation of the materials transported in solution.

Each geomorphic agent functions in a way peculiar to it and as a result produces erosional and depositional land-forms which are characteristic of that agent. It is therefore possible to recognize the origin of land-forms although the processes by which they were formed may have ceased to operate ; it is because 'geomorphic processes leave their imprint upon land-forms'.

The structure of the crust acts as a control in the formation of land-forms. Folds, faults, unconformities, differences in the hardness and susceptibility of rock units to decomposition and disintegration provide a frame-work within which the geomorphic agents function. The resulting physiographic features are always influenced by the structure of the underlying rock. Thus the combination of a particular structure being eroded primarily by one geomorphic agent will result in a predictable land-form.

Cycle of erosion. The concept of cycle of erosion was formulated by William Morris Davis. According to him, in a cycle of erosion the surface features undergo changes as a result of the processes acting upon them. These changes are systematic and follow each other in a regular sequence. These sequences are termed stages. 'Davis has recognized four stages'.

(*a*) Initial stage   (*b*) Youth stage
(*c*) Stage of maturity   (*d*) Old-stage.

However, most commonly, only youth, maturity and old-stages are used in the study of topographies produced by different geomorphic agents.

In the initial-stage the land-form is generally even and is raised high above the sea-level, where small and slow changes occur. Most of these features are erosional in origin.

However, vast changes occur by the time the landscape attains maturity.

In the old stage again landscape evolution slows down.

Different geomorphic agents have their characteristic cycles of erosion. Thus we have fluvial cycle of erosion, glacial cycle of erosion, aeolian cycle, karst cycle and marine cycle of erosion.

The cycle of erosion, operates through the three stages until the initial relief is almost fully reduced and the surface is again levelled. Most of the cycle of erosion do not reach the final stage as some time during their operation either climatic or tectonic disturbances take place. An incomplete or partial cycle results from this. Topography returns to a youthful stage. This phenomenon is known as '*rejuvenation*' whereby a mature topography becomes young.

The various land-forms and other topographic features associated with the different geomorphic agents owe their origin to the geological actions like erosion, transportation and deposition of the concerned geomorphic agents and are as follows :

1. Geological-action of river. This phenomenon, which is associated with the geological action of river is usually known as the fluvial cycle of erosion, or of the normal cycle of erosion.

Erosion. The erosion caused by the running water is of two types :

(*i*) Mechanical erosion, and (*ii*) Chemical erosion.

(*i*) **Mechnical erosion.** It is because of the physical forces associated with the running water and it takes place in four distinct manners like :

(*a*) *Hydraulic action.* Forces inherent in the flow of running water, can do a great deal of erosion of the bank and the bed-rock. It is mostly due to surface relief, *i.e.*, gradient.

(*b*) *Abrasion.* The materials which are being carried away by the running water acts as tools of destruction, and during their transportation, because of their rubbing against the surface of the bed-rock, they bring about a scraping of the surface. This process of erosion is also known as·'*Corrasion*'.

(*c*) *Attrition.* Materials during their transit often collide among themselves and in turn get themselves teared and this is the process, through which big boulders are gradually reduced in size and finally reach the size-grade of sand and silt.

(*d*) *Cavitation.* This is because of the presence of the air bubbles which create a whirling action at the time of penetration of water through the existing pores and fissures, and the small sard particles along with the air bubbles play a major role in widening the cavities

Factors Which Help Mechanical Erosion

1. Hydraulic gradient.

2. Climate—which determines precipitation and finally volume and velocity of water.

3. Nature of the bed-rock, whether it is hard or soft ; whether the layering or jointings of the bed rock are parallel to the flow of the water or are perpendicular to the same ; whether the bed rocks are igneons, sedimentary or metamorphic rocks.

4. Hardness of the transported materials.

(*ii*) **Chemical erosion.** It is also known as solution or 'Corrosion', during which process the materials get dissolved in the water of the river and are transported in solution.

Factors

(1) Dissolving action of water due to the presence of carbon-dioxide.

(2) Solubility of the river bed.

Transportation. There are two methods of transportation :

(*a*) Mechanical, and   (*b*) Chemical.

(*a*) **Mechanical transportation.** It takes place in three ways :

(*i*) Suspension (floating).

(*ii*) Traction (by creeping and rolling).

(*iii*) Saltation (through lifts and falls of materials).

These processes are aided with the following factors :

(*a*) Velocity of the river.

(*b*) Nature of the river-current.

(*c*) Density of the rock-material to be transported.

(*b*) **Chemical transportation.** It is through the process of solution, usually in the form of carbonates, sulphates of calcium, sodium, potassium, magnesium etc.

Deposition. It is the last geological action by the river, whereby materials transported get accumulated in an appropriate site where the following factors play major roles :

(*i*) Decrease in velocity of the transporting medium.

(*ii*) Decrease in slope.

(*iii*) Decrease in volume.

(*iv*) Change in channels.

(*v*) Chemical precipitation.

The main features of the '*fluvial cycle of erosion*' are as follows :

The cycle begins on a recently uplifted landmass. It is initiated through the drainage system working on it.

1. During the '*Initial Stage*' a river is formed and it involves some of the characteristic processes like :

(*i*) Channel deepening due to bed-scouring,

(*ii*) Pot-hole drilling (by whirling current action).

(*iii*) Tributaries are fast developed.

(*iv*) Headward erosion is maximum.

(*v*) Waterfalls, gorges and canyons are formed.

2. Youth-stage. River system is to some extent established. River capture or river piracy takes place, in which case, one river reaches the course of another one and if the course of the second river is diverted because of the greater gradient of the earlier river, this phenemenon will be known as *river-piracy*. The point where the course of the second river is diverted is known as '*elbow of capture*'. The captured river is called the '*misfit-river*' and the abandoned part of the channel, through which no water flows is the 'wind-gap'.

In this stage, there is conspicuous formation of V-shaped valleys.

3. Mature-stage. In this stage there is maximum erosion through lateral-cutting. The landmass is fully dissected ; ridges and valleys develop strikingly.

As a result of heavy erosion and deposition this stage is comprised of a large number of landforms such as hog-backs, cuestas, plateau tables, meanders, ox-bow lakes, terraces, alluvial fans etc.

4. Old-stage. Down-cutting stops at the *base level of erosion*, which is the mean sea level produced inland. Frequently the river is over-flooded and builds up *flood-plains* on both sides. The initial irregular surface has become practically flat at this stage. The river is mostly engaged in depositing and does little of erosion and transportation. The topography is characterised by a set of distributional features which comprise delta, distributaries etc.

Features and typical-landforms associated with an ideal fluvial cycle in a humid region :

(*i*) **Strath.** When the width of the valley is greater than the width of the river, the valley is called a *strath*.

(*ii*) **Knick-point.** During the process of regrading, there is a more or less marked change of slope at the point of intersection of newly graded profile with the older one, and it is known as the knick-point.

(*iii*) **Bad-land.** Due to pronounced erosion by running water the areas are intricately traversed by gullies, which mostly develop on argillaceous rocks.

(*iv*) **Escarpment.** It is a steep slope resulted from differential weathering of rocks.

(*v*) **Cuesta.** In a region of sub-horizontal beds, a gentle slope is developed along the gentle dips of strata, such a landscape is known as cuesta.

(*vi*) **Mesa.** An isolated table-land area with steep sides.

(*vii*) **Butte.** With continued erosion of the sides a mesa is reduced to a smaller flat-topped hill, known as '*butte*'.

(*viii*) **Hog-back.** It is a cuesta, in which the dip slope and scrap slope are both approximately 45°.

(*ix*) **Braided-river.** In this case distributaries or branches develop in large number in a region of flatness. They are commonly formed where the amount of load is excessive and the stream is incapable of transporting all of it. The coarser fractions of the load tend to form islands in the centre of the stream, which breaks up paths around them. Thus a braided river is formed.

(*x*) **Peneplains.** The peneplains are formed in the old stages of rivers and are the plain lands produced by the river.

(*xi*) **Monadnock.** Sometimes some mounds or small hillocks of hard rock persists on the peneplains and are knows as *monadnocks.*

(*xii*) **Natural levee.** On the flood-plains, long depositional ridges extending parallel to the river are found, which are known as Natural-levees.

Features Found in Arid Regions

(*i*) **Alluvial fans and cones.** On descending to the plains from the hills the velocity of a river and the carrying capacity are reduced. At this point the river-sheds a large amount of load which assumes a fan or concial shape.

In the lower parts, many fans join laterally to produce a '*bajada or piedmont plain.*'

(*ii*) **Pediment.** It is a plain of eroded bed rock in an arid region developed between mountain and basin areas. Pediments converge to form pediplains.

(*iii*) **Bornhardts.** If pediments occur on both sides of a mountain range, they gradually converge. Eventually the range is reduced to a very broad dome with the slope equal to that of the pediments on either side of the crest that consists of a narrow ridge of small scattered domes. These residual hills are known as Bornhardts.

(*iv*) **Playas.** In deserts that consists of basins enclosed by mountain ranges the drainage is towards the centre of the basin from all margins When there is sufficient water, this plain is covered by a broad shallow lake called playa.

(*v*) **Wadies.** These are channels formed during rains in desert or arid regions.

(*vi*) **Inselbergs.** These are isolated mounds rising above the general level of a pediment. These are equivalents to monadnocks in humid regions.

Depositional Features

(*vii*) **Deltas.** These are submerged equivalents of alluvial fans.

(*viii*) **Sloughs.** Depressions on the flood-plains of meandering rivers, which are excavated during floods due to the tendency of the overflowing water to follow a short-course.

Important Terminologies Associated with the Actions of this Geomorphic Agent :

(*a*) **Effluent and influent river.** In case of effluent rivers, the regional water-table lies near the earth's surface, as a result the river or stream is fed ; whereas in case of influent rivers, the water-table is located at a great depth, a part of runoff is scheduled to percolate downwards.

(*b*) **River pattern :**

(*i*) **Antecedant.** Rivers existing before the surface relief was impressed upon the area.

(*ii*) **Consequent.** The flow of the river occurs as a consequence to the existing surface relief.

(*iii*) **Subsequent.** The river which joins the consequent river arising later as erosion proceeds.

(*iv*) **Insequent.** It displays no reason for its particular course, such as that upon homogeneous terrain.

(*v*) **Obsequent.** Here the river drains in the opposite direction to the original consequent-river.

(*vi*) **Resequent.** It drains in the same direction as the original consequent, but at a lower topographical level.

(*vii*) **Super-imposed.** At some places, old rocks may be covered under a sheet of new deposits. Any river developed on such an area will follow the surface relief of the overlying cover and will not have any relation with the older rocks lying below. Gradual erosion removes the overlying cover and the river flows on the older rocks below. Here, the river is said to be super-imposed on the older rocks below.

(*c*) **Drainage pattern.** The joining of the tributaries with the master stream produces a pattern termed drainage pattern. The common drainage patterns are :

(*i*) **Dendritic.** Which is characterised irregular branching of tributary streams in a similar pattern as that of a tree's branches.

(*ii*) **Parallel pattern.** Develops on steep slopes where the tributaries and the master stream flow parallel to each other.

(*iii*) **Trellis pattern.** It develops in a topography created on a folded structure of synclines and anticlines, faults or joints etc.

(*iv*) **Radial pattern.** It consists of drainage lines radiating from a central part as on a dome.

Geomorphic Features Produced Due to Wind-Action   59

the wind serve as tools of destruction and whether they, on some
rock-surface they bring about a scraping — abrasion.

CHAPTER 15

GEOMORPHIC FEATURES PRODUCED DUE TO – WIND-ACTION

Wind-action is conspicuous in semi-arid and arid regions, but it is particularly strong in deserts. Aeolian topography is created by the geological faction of wind, which can conveniently be divided into the following three stages :

(*a*) Erosion, (*b*) Transportation, (*c*) Deposition.

(*a*) **Erosion.** The wind accompolishes erosion by three means :

(*i*) Deflation.   (*ii*) Abrasion.   (*iii*) Attrition.

(*i*) **Deflation.** Deflation is the process of removal of loose soil or rock-particles, along the course of the blowing wind. This process operates well in dry regions with little or no rainfall.

(*ii*) **Abrasion.** Abrasion is the sand blast action of wind with sand against the rocks. The loose particles that are blown away by

(*i*)   (*ii*)   (*iii*)

Fig. 15·1.   Showing stages of development of ventifacts due to wind-abrasion.

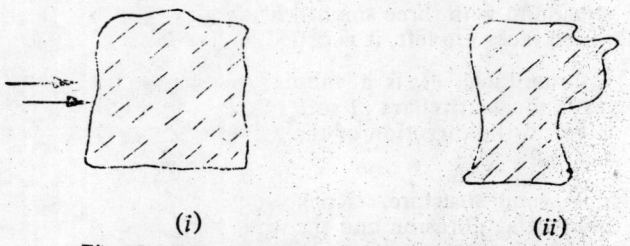

(*i*)   (*ii*)

Fig. 15·2.   Formation of pedestal rock due to wind-abrasion.

the wind serve as tools of destruction and when they move on some rock-surface they bring about a scraping of the surface.

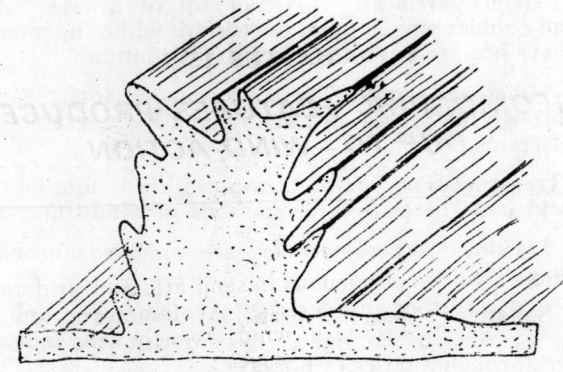

Fig. 15·3. Yardang.

(*iii*) **Attrition.** Attrition is the grinding action. While on transit wind born particles often collide with one another. Such mutual collision brings about a further grinding of the particles.

Important Erosional Features and Associated Landforms

(*i*) **Hamada.** Due to deflation, when the loose particles are swept away, only the hard mantle is left behind which is known as *Hamada*.

(*ii*) **Yardang.** A grooved or furrowed topographic form produced by wind abrasion, which is elongated in the direction of prevailing winds and is usually strongly under cut, is known as Yardang.

(*iii*) **Pedestal rock.** A wide rock-cap standing on a slender rock column, produced because of the wind-abrasion, is known as a pedestal rock.

(*iv*) **Ventifacts.** These are pebbles faceted by the abrasive effects of wind-blown sand. Ventifacts with one smooth surface is called *Einkanters* and with three smooth faces as *Dreikanter* ; when only two abraded faces are left, it is called *zweikanter*.

(*v*) **Mushrom-table.** It is a tabular mass of more resistant rock resting on under-cut pillars of softer material. They are very often elongated in the direction of the prevailing wind and are also known as '*Zeugen*'.

(*vi*) **Honey-comb structure.** Rocks consisting of hard and soft parts get differential abrasion and the resulting feature is known as honey-comb structure.

(*vii*) **Blow-outs.** These are broad-shallow caves in hills, broad-shallow depression in deserts.

(*viii*) **Desert pavement.** It is made up of a layer of residual pebbles and cobbles strewn upon the surface while intervening finer particles have been removed as a result of deflation.

(*xi*) **Millet-seed sands.** These are rounded desert sand grains, resulted through the process of attrition and have resemblance with millet seed grains.

(*b*) **Transportation.** Wind-transportation is totally dependent on wind-velocity. There are three methods of wind-transportation :

(*i*) **Traction.** Where particles are removed through rolling and creeping.

(*ii*) **Saltation.** Here the particles, which are too heavy to remain in suspension and lighter to be transported in traction, are transported through a series of bounces.

(*iii*) **Suspension.** Very light particles like dust and cloud, smoke etc. move with the wind quickly but settle very slowly, remain in suspension in the air.

(*c*) **Deposition.** Wind-formed deposits are called aeolian deposits. Wind is an excellent agent for sorting of materials according to their size, shape or weight. Pebbles and boulders cannot be carried away and are left back to form *lag deposits*. The clayey and silty fractions are deposited as *loess*, which does not show any stratification. Wind deposits take two general forms as :

(*i*) Sheets,   (*ii*) Piles.

'*Sheet deposits*' are the dust deposits laid down on large area. '*Piles deposits*' include the various types of dunes which accumulate from sand and silt carried in saltation.

Depositional Features :

(*i*) **Sand hill.** Mounds of sand whose surface is irregular is called sand hill.

(*ii*) **Sand dune.** When the mound is in the form of a round hillock or a ridge with a crest, it is called a sand dune. In structure a dune has a gentle slope towards the wind-ward side and a steep-face towards the lee-side. The shape of a dune is controlled by

(*a*) amount of sand supply,

(*a*) wind-velocity,

(*c*) constancy of wind direction, and

(*d*) amount and distribution of vegetative cover.

Types of Sand-Dune

(*i*) **Barchan.** Barchan are the crescentic shaped dunes with the points or wings directed downwind.

(*ii*) **Seif.** It is similar to barchan except one wing is missing, caused by an occasional shift in wind direction.

(*iii*) **Transverse dune.** Elongated dunes form at right angles to the prevailing wind.

Fig. 15·4A

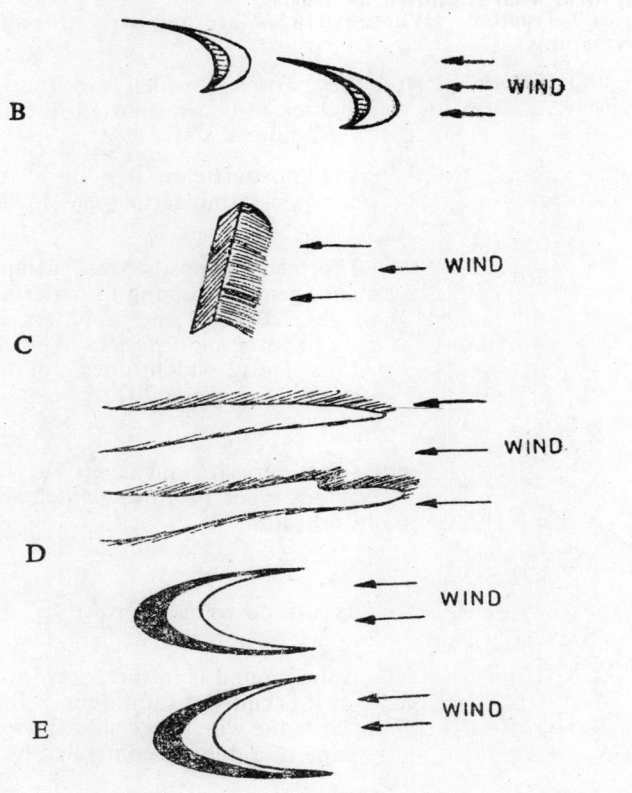

Fig. 15·4. (A) These are, just the barchans, (B) One wing of barchan is missing here. (C) Transverse dunes. (D) Logitudinal dunes. (E) Parabolic dune.

(*iv*) **Fore dune.**   Ridge-like deposits of wind borne sand formed along the coast of sea or lakes.

(*v*) **Longitudinal dune.**   Elongated ridges of sand found to lie parallel to the direction of blowing wind.

(*vi*) **Parabolic dune.**   These are of parabolic shape, their horns point towards the direction opposite to that of the blowing-wind.

(*vii*) **Whale back dune.**   It is a very large longitudinal dune with flat tops, on which barchans or seifs may occur.

The space between the dunes is called '*Gassis*' and the water which is available in shallow wells and support vegetation in desert areas, form what is known as '*Oasis*'.

CHAPTER 16

LANDFORMS MADE BY GLACIERS

Davis suggested that a glacial topography is a climatic accident that happens to normal cycle of erosion, *i.e.*, climate gets very cold and the river freezes. Instead of rivers of water there are rivers of ice, called glaciers, operating as the main geomorphic agent. The geological action of glaciers, *i.e.*, erosion, transportation and deposition together constitute, what is known as glaciation.

Formation of glaciers. Under the influence of pressure and moisture, the snow flakes change into a granular ice mass, known as névé in French and 'firn' in German. When the ice becomes so thick that the lower layers become plastic, outward or downhill flow commences and an active glacier comes into being.

Type of glaciers. There are three major types of glaciers :

(*i*) The valley glaciers (Mountain glaciers or Alpine glaciers).

(*ii*) Piedmont glacier—these are intermediate in form and origin, between valley glacier and ice-sheet. They are formed at the foot of the mountains.

(*iii*) **Ice sheet.** These are huge covers of ice and are also known as *'Continental glacier'*.

Accordingly, there are two types of glacial topography, one related to the valley glaciation and the other to the continental glaciation.

Topography of valley-glaciation. There are two sets of features resulting from glacial erosion and glacial deposition, (since the rock-wastes and other loads are carried frozen within the ice during their transportation, in case of glaciers).

Erosion. Erosion by glaciers take place due to (*i*) *plucking*, (*ii*) *rasping* and (*iii*) *avalanching*.

Plucking is also known as frost-wedging or quarrying. During the summer months the surface parts of a glacier may partially melt. This melt water or rain water seeps down along the sides of the ice mass, finding its way into the cracks and fractures in the rocks along the edges and at the head of the glacier. At night or when the temperatures drop this water freezes. It breaks up the rock by frost-action, and with the movement of the glacier they are frozen in suspension in the ice.

Rasping is the term used to describe the scraping or abrasion by glacial action. The front edge of glaciers function as bulldozer, pushing and scraping the ground in front of the ice.

Avalanching is the process of mass-wasting. Along the margins of a valley-glacier the valley sides are scrapped and blocks are broken off which become frozen into the ice and are carried away. This leads to under-cutting of the sides of the valley and pave the ground for slumping, sliding and debris avalanching which bring great quantities of rock-waste on to the top surface of the glacier.

Features of Erosion

1. Cirques. These are circular depressions formed by plucking and grinding on the upper parts of the mountain-slopes. These are also known as *'corries or amphitheatre'*.

2. Arete. This name is applied to the sharp ridges produced by glacial erosion. Where two cirque-walls intersect from opposite sides, a jagged, knife-like ridge, called an *'arete'* results. It is also known as *'comb'* or *'serrate-ridge'*.

3. Horn. Where three or more cirques grow together, a sharp-pointed peak is formed by the intersection of the aretes. Such peaks are known as 'horns'.

4. Col. Where opposed cirques have intersected deeply, a pass or notch, called a *'col'* is formed.

5. Glacial-trough. Glacier flow constantly deepens and widens its channel so that after the ice has finally disappeared there remains a deep, steep walled, 'U'-shaped valley, known as glacial trough.

6. Hanging valley. Tributary glaciers also carve 'U'-shaped troughs. But they are smaller in cross-section, with floors lying high above the floor-level of the main-trough, *i.e.*, main glacial valley. Such valleys are called hanging-valleys.

7. Fiords. When the floor of a glacial trough open to the sea lies below sea-level, the sea-water will enter as the ice-front recedes, producing a narrow estuary, known as a *'fiord'* or *fjords'*.

8. Tarns. The bed-rock is not always evenly excavated under a glacier, so that floors of troughs and cirques may contain rock-basin and rock-steps. Cirques and upper parts of troughs thus are occupied by small lakes, called tarns.

Depositional features of valley-glaciers

Deposition by a glacier takes place when the ice begins to melt and the glacier slows down and vanishes, losing its transporting power. The unstratified, unsorted debris dropped more or less in a random fashion by glaciers form deposits known as *morains*. Three types of moraines are known, *lateral, medial or median*, and *terminal or end*. These three types are differentiated on the basis of their location in the valley.

(*a*) **Lateral-moraine.** Deposits of ridge-like pattern formed along the margins of the glaciated valley are known as lateral moraines.

(*b*) **Medial-moraine.** It results due to coalescence of two lateral moraines, where two ice streams join.

(*c*) **Terminal-moraine.** These are accumulation of rock-debris at the terminus of a glacier.

(*d*) **Recessional moraine.** Where glacier retreats in a halting manner, a series of concentric moraines is formed, known as '*recessional moraines*'.

Topography of 'continental glaciation'. Like valley glaciers, the continental glaciers proved to be highly effective eroding agent. But continental glacier erodes only by plucking and rasping methods, but erosion process, like avalanching is absent in case of continental glaciers.

Features of Erosion :

(*i*) **Striations.** The slowly moving ice scraped and ground away much solid bed-rock. Left behind were smoothly rounded rock-masses bearing countless minute abrasion marks, scratches, called *striations*.

(*ii*) **Roches moutonnees.** They consist of asymmetrical mounds of rock of varying size, with a gradual smooth abraded slope on one side and a steeper rougher slope on the other. The 'stoss side', *i.e.*, the side from which the ice was approaching is characteristically smoothly rounded and the other side, *i.e.*, the 'lee side' where the ice plucked out angular joint blocks, is irregular and blocky. They are also known as '*sheep-rocks*'.

(*iii*) **Crag and tail.** Sometimes very hard rocks like volcanic plugs offer great resistance to the ice-flow and stand as pillars in the glaciated valley. These structure are called crags and the lee side which is sloping in this case is the tail.

Depositional features. The term glacial drift includes all varieties of rock debris deposited in close association with glaciers. These deposits may be classified into two groups :

(1) Stratified drift, consists of layers of sorted and stratified clays, silts, sands etc. deposited by the meltwater streams and are also known as Glacio-fluvial deposits.

(2) Till. It is a heterogeneous mixture of rock fragments ranging in size from clay to boulder which are unsorted and unstratified. A consolidated till is called 'tillite'. The various depositional features are as follows :

(a) **Drumlin.** It consists of glacial till, which is a low mound of clay containing cores of bed-rocks. Uphill sides are blunt and down-hill sides are smooth and gently sloping. The long axis of each drumlin parallels the direction of ice-movement and thus serve as indicators of direction of ice-movement.

(b) **Basket of egg-topography.** The drumlins commonly occur in groups of swarms, which may number in the hundreds ; the topography produced by them is peculiar and is known as basket of egg-topography.

(c) **Ground moraine.** Between moraines, the surface over-ridden by the ice is overspread by a cover of glacial till, known as ground moraine. Thus it is the sheet of debris left after a steady retreat of ice.

Glacio-Fluvial Deposits :

(i) **Outwash plain.** It is also known as overwash plain. Glacial streams carry a huge quantity of rock-debris and then form fan-like plains beyond the terminus of glaciers. These are stratified. When they occur on valley floors such outwash plains are called 'valley trains'.

(ii) **Kames or kame terraces.** These are formed on the top surface of a glacier where the surficial melt-waters wash sediments from the top into depressions. As the ice melts the material that formerly filled depressions on top of the glacier is dropped and makes small hills, which are more or less flat-topped and are known as kames. Terraces, called kame terraces, are built in this way.

(iii) **Eskers.** These are winding steep-sided ridge-like features built of stream borne drift. These are also known as *Osser or Oss.*

(iv) **Erratics.** These are stray boulders of rocks which have undergone a prolonged glacial transport and have subsequently been deposited in an area, where the country rocks are of distinctly different types. At times they are delicately balanced upon glaciated bed rock, and are called *poking* or *logging-stone.*

(v) **Kettles.** Drifts occurring in the vicinity of a glacier and particularly those lying near about the ice-terminus are ordinarily found to contain a number of depressions, some of which may give itse to lakes or swamps. Such hollows are known as kettles.

(*vi*) **Varves.** These are layered clays alternating with coarser and finer sediments.

Other Important-Features Associated with Glacier :

(*i*) **Nivation.** It is the process of quarrying of rocks mostly by frost action.

(*ii*) **Ablation.** It includes processes both evaporation and the melting of snow and ice.

(*iii*) **Calving.** Within fiords, glaciers come in contact with marine water and blocks of ice are found to break-off from the mass of the glacier. This process of wastage of glacier is known as calving.

(*iv*) **Serace.** Similar to a waterfall in a river, in a steeper section of the valley, the glacier is broken up into rugged ice-pinnacles and is known as *serace*.

(*v*) **Iceberg.** These are floating ice-hill on the sea-water.

(*vi*) **Crevasses.** These are cracks formed due to differential movement within the mass of glacier. In German, they are known as *Bergschrund*.

(*vii*) **Nunatak.** A rock-mass which projects through an ice-sheet, generally found at the margins of a sheet where the ice is thinnest, is known as *nunatak*.

CHAPTER 17

LANDFORMS MADE BY THE ACTION OF UNDERGROUND-WATER

Ground-water is that part of the sub-surface water which fully saturates the porespaces of the bed-rock and regolith. Surface water, after infiltration, becomes a part of the underground water and it is also known as sub-surface water. The ground-water occupies the *saturated zone*, above it is the zone of aeration. The upper surface of the saturated zone is the *Water-table*.

Mechanical process of erosion is absolutely insignificant in case of underground water. It brings about erosion only through chemical processes, *i.e.*, by the solution action of the underground water. This process is particularly effective in regions of soluble rocks like limestone, dolomite etc..

Topography developed due to the action of ground-water is known as '*Karst-topography*', after the occurrence of the typical limestone topography in the Karst area of Yugoslavia. Karst topographic features develop on both over and under the ground.

Karst topography develops very well where the following two essential conditions are fulfilled :

(*i*) the soluble rocks are located near or at the earth's surface.

(*ii*) the rocks are dense, highly jointed, and thin-bedded.

Two secondary conditions also favour this development which are : (*a*) the presence of a deeply entrenched valley of a master stream, and (*b*) a moderate amount of rainfall.

Erosional Features :

(*i*) **Lapies.** The leaching action of the ground-water as it passes through the limestone region, produces a highly rugged topography. The ground water may enlarge the joints of the limestone into a conjugate pattern of clefts and ridges, this surface is called a *limestone pavement* or lapies surface. The clefts in such a pavement are called grikes, and the ridges clints.

(*ii*) **Sink.** It is a conical depression in the limestone, which may be several metres in diameter. It is also called 'doline'.

(*iii*) **Uvala.** An uvala is a very large elongated depression formed by the convergence of two or more sink-holes.

(*iv*) **Caverns.** These are hollows underground with their roofs intact.

(*v*) **Galleries.** These are horizontal linking passages of connecting caverns.

(*vi*) **Shaft.** These are vertical or inclined linking passages of connecting caverns.

(*vii*) Some characteristic underground or surface features occur in limestone caverns. These features including *stalactites* and *stalagmites*, which together constitute what is known as '*drip-stone*'.

Stalactites are columns of limestone that hang from the ceiling downwards and stalagmites are the columnar features which rise up vertically from the floor of the cavern.

(*viii*) **Polje.** Larger depressions in the landscape (covering tens of square kilometres) are known as poljes and are characterised by steep sides and flat floors If the water-table is high enough lakes may form and they are known as '*polje lakes*'.

(*ix*) **Hums or pepino hills.** These are small residual hills found on the floors of polje.

(*x*) **Stylolite.** It is an irregular suture-like boundary developed in some limestones, in which less soluble portions of any two consecutive beds project into each other, and thus form a zig-zag line of junction of the beds concerned.

(*xi*) **Natural bridges.** It represents the remnant of the roof of a natural tunnel or subterranean cut off.

(*xii*) **Swallow holes.** Sometimes sink-holes become so numerous that the sides begin to touch one another. Surface drainage becomes limited to short sinking creeks, those that disappear into the ground. Along some such streams there are small holes where water swirls into small openings leading into caverns. Such holes are called *swallow holes*.

(*xiii*) **Blind valleys.** These are valleys that lead into a hill side or gradually lose the characteristics of a valley as the water from their streams is lost to sub-surface channels.

(*xiv*) **Karst valley.** It is a very deep valley, formed by the solution process and occurs in limestone rocks.

Depositional Features :

(a) **Geode.** It is a cavity in a rock lined with quartz crystals projecting towards the centre.

(b) **Sinter.** Deposits of silica or calcium carbonates by ground water are known as silicious or calcareous sinters.

(c) **Kanker.** These are loose gravel and alluvium which have been subjected to cementation and compaction.

Karst cycle. It consists of four stages, of which the youthful stage is characterised by progressive expansion of the underground drainage. The mature stage displays lakes, uvalas, and caverns. The late maturity stage shows decline of karst features. The old stage reveals the reappearance of streams and entrenched valleys on the surface.

CHAPTER 18

LAKES

Lakes are bodies of water, either fresh or saline, in natural depressions on the surface of the earth. As we know, the geological agents individually as well as collectively tend to reduce the surface of the earth to a continuous and gradual slope. In doing so, sometimes large depressions are formed on the land surface, which when filled up with water are described as lakes. They range in size from a pond to larger ones of hundreds square miles in area.

Lakes are distinguished from swamps since the lakes commonly occur above the mean-sea level and the swamps are low-lying lands where the water-table has just reached the land-surface. Basins are differentiated from lakes as they always have their bottoms below the water-table.

Origin of Lakes

1. By river action :

(a) Lakes may be formed by meandering rivers, which are called as ox-bow, horse-shoe or cut-off lakes.

(b) Rivers may form lake basin by erosion at the foot-hill region as a result of the impact of waterfall.

(c) Tributary river by forming bar of sediments across the main-river may block it to form lake.

(d) Lakes may be found to occur in depressed areas in dried up river beds.

(e) Lakes may be formed by rock-fall or land slide, when the river is blocked by the landslide across its valley.

2. By tectonic movements :

(a) Tectonic depressions relative to the surrounding are responsible for the largest of the world's lakes like Caspian sea, Baikal, Dead sea, Titicaca etc.

(b) Folding as well as differential faulting like tear faults or thrusts across the pre-existing river valley.

(c) Lakes may also result due to earthquakes.

3. Due to volcanic activities :

(*a*) Lakes are sometimes formed on the craters and calderas of extinct or dormant volcanoes.

(*b*) Lava flows forming barriers across pre-existing valleys.

4. Due to glacial actions :

(*a*) Piling up of morainic matter across their valleys causing the formation of lake.

(*b*) Kettle-holes left by melting of masses of stagnant ice.

(*c*) Valleys obstructed by glacial drift.

5. Marine action. Sometimes sea-waves build bars across the coast or the mouth of a river, thus converting the lake water into small lagoons.

6. Wind action. Due to extreme grade of deflation hollows are excavated in arid regions to a depth where an adequate supply of ground is available, thus forming aeolian lakes.

7. By the impact of large meteorites lakes may be formed.

8. Organic activity such as growth of coral reefs leads to the development of lagoons with the emergence of atolls.

9. Ground water action on soluble rocks leads to the collapse of the roof-rock giving rise to lakes, which are known as Poljee.

10. By constructing dams, artificial lake can be created.

Geological action. Excepting an inconspicuous degree of erosion brought about by lake-water the geological action of lakes are mostly concerned with the deposition of sediments as it becomes a depositional site for the detritus carried by the streams and rivers feeding the lakes.

Lakes in India

(A) Peninsular India :

1. Coastal lakes : (*a*) Chilika lake, (*b*) Pulicat lake, and (*c*) Kayal (Kerala).

2. Lonar lake. In Buldana area of Maharashtra.

3. Sambar lake. In Rajasthan

4. Dhands. Small lakes of aeolian origin.

5. Runn of Kutch. In Gujarat.

(B) Extra peninsula :

1. Lakes of Kashmir. Walur, Dal are fresh water lakes while Pangkong, Salt lake, Tsomoriri are saline.

2. The lakes of Kumaon which include Naini Tal, Bhim Tal etc., are thought to be of tectonic origin.

CHAPTER 19

IMPORTANT FEATURES RELATING TO OCEANS AND SEAS

1. **Continental shelf.** It is that part of the sea-floor adjoining a land mass over which the maximum depth of sea-water is 200 metres. It is a belt of shallow water and is mostly covered by sediment. The average width is about 70 km and the mean slope is less than one degree. The continental shelves cover about 7.5 percent of the total area of the oceans and 18 percent of the land. About 20% of the world production of oil and gas comes from them.

2. **Continental slope.** The continental slope joins the shelf to the deep ocean floor. The average gradient of the slope is about four degrees. It is a steeper zone that gives place to the continental rise which in its own turn is succeeded at far greater depths by the abyssal plain. The depth of sea-water in this zone ranges from 200 metres to 1000 metres. It is the margin of continental landmasses.

3. **Continental rise.** It is a prism of thickened sediments located below the surface of the continental slope. The material of the rise has been derived from the shelf and slope. All the continental rises are separated from the shelf by the continental slope.

4. **Abyssal plains.** It is the zone of deep sea-floor and the depth of sea-water here ranges from 1,000 metres to 4,000 metres. It is characterised by a slimy, mud-like deposit called 'ooze' and red-clay.

An associated feature with the oceans or seas is :

(i) **The coral-reef.** There are three types of coral-reef as

(a) *Fringing reef.* Which are built as platforms attached to shore, i.e., close to the main islands or volcanic cones.

(b) *Barrier reef.* Which are built away from the main land and separated from the main land by a lagoon ; in other words, they enclose a body of water between the mainland and themselves.

(c) *Atoll.* These are more or less circular coral reefs, enclosing a lagoon but without any land inside.

(*ii*) **Spits.** Near about the shore the rock debris sometimes are heaped up in the form of ridge running more or less parallel to the shore. These ridges are in the long run visible above the sea-level, and are known as spits.

(*iii*) **Bars.** If the spits connect themselves with the coastal tract, they are described as 'bars'.

PART III

STRUCTURAL
AND
FIELD GEOLOGY

CHAPTER 20

DIP AND STRIKE, CLINOMETER-COMPASS AND ITS USES

As we know, most sedimentary rocks were originally deposited on flat or very gently inclined surfaces. Sometimes, of course, as a result of special circumstances, certain beds may have started with an initial inclination, as in current bedding ; but where we find great thickness of strata tilted into conspicuously inclined positions, it is to be borne in mind in general that the beds have been tilted by movements that occurred after their deposition. The attitude of an inclined bed has got two important components as

    1. Dip   2. Strike.

In specifying the attitude of any inclined bed, *i.e.*, to define its position in space, it is required to determine the dip and strike of the concerned bed.

1. Dip. What it is ?

It is essentially an angle of inclination of the bed. It is defined as the amount of inclination of a bed with respect to a horizontal plane ; measured on a vertical plane lying at right angles to the strike of the bedding.

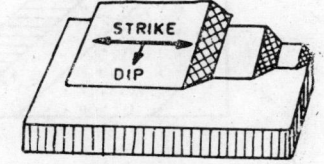

Fig. 20·I

The dip of a bed has got two components, like direction as well as magnitude. So the dip of a bed is a vector quantity. The amount of dip is the angle which varies from '0°' to '90°' according to the disposition of the bed. The direction of dip is the geographical direction, along which a bed has maximum slope.

In case of horizontal beds, the dip is 'zero' degree and for a vertical bed the dip is '90°'. Accordingly, the symbolic representation of a horizontal and vertical bed in a map is also different. which may be seen from the following figures :

A.   Horizontal bed :

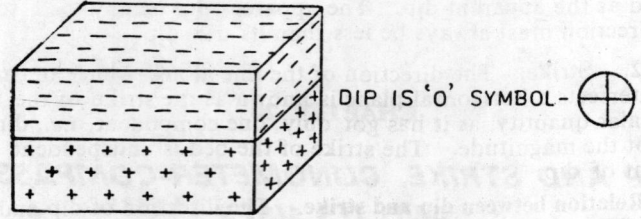

DIP IS '0' SYMBOL—

Fig. 20·2

B.   Vertical bed :

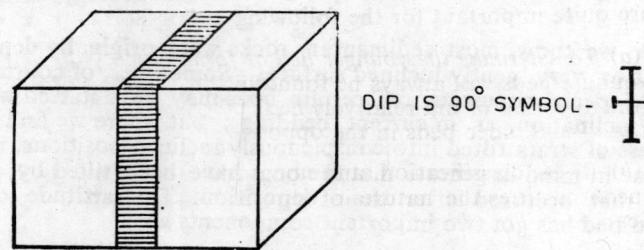

DIP IS 90° SYMBOL—

Fig. 20·3

C.   Inclined bed :

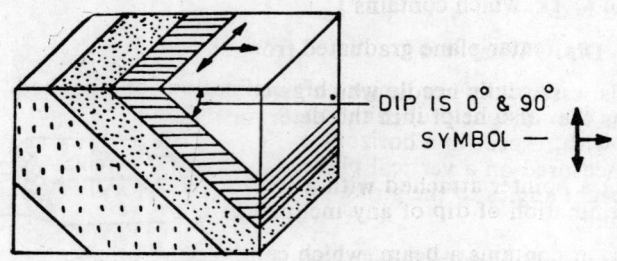

DIP IS 0° & 90°
SYMBOL —

Fig. 20·4

Types of dip.   There are two types of dip as

(*i*) True dip   (*ii*) Apparent dip.

(*i*) **True dip.**   It is the maximum amount of slope along a line perpendicular to the strike, in other words, it is the maximum slope with respect to the horizon.

It may also be stated as the geographical direction along which the line of quickest descent slopes down.

(*ii*) **Apparent dip.** Along any direction other than that of the true dip, the gradient is scheduled to be much less and therefore it is defined as the apparent dip. The apparent dip of any bed towards any direction must always be less than its true dip.

2. Strike. The direction of the line along which an inclined bed intersects a horizontal plane is known as the strike of the bed. It is a scalar quantity, as it has got only one component, *i.e.*, direction but not the magnitude. The strike of the bed is independent of its amount of dip.

Relation between dip and strike. The direction of dip and strike of any inclined bed must lie at right angles to each other. True dip is in the direction along a perpendicular to the strike.

Importance of strike and dip. In structural geology, strike and dip are quite important for the following purposes :

(*a*) *To determine the younger bed or formation.* It is well known that younger beds will always be found in the direction of dip. If we go in the direction of dip relatively beds of younger age will be found to out crop and older beds in the opposite direction.

(*b*) In the classification, and nomenclature of folds, faults, joints and unconformities the nature of dip and strike is of paramount significance.

Thus the attitude, which refers to the three-dimensional orienta tion of some geological structures, is defined by their dip and strike.

Clinometer-compass. It is essentially an equipment for geological field work, which contains **;**

(*i*) a circular plane graduated from 0° to 360° ;

(*ii*) a magnetic needle which gives the direction of north and south and it is also helpful in the determination of the strike of any inclined bed ;

(*iii*) a pointer attached with a freely oscillating pendulum for the determination of dip of any inclined bed, and their direction ;

(*iv*) it contains a beam, which can be only rotated through an angle of 90°, is having a small hole at one end and a vertical window at the other, with a vertical separator. It is used to take bearing (which is 'back bearing') at the time of determining one's location in the field.

Thus clinometer compass is used for determining the dip and strike of beds in the field, for surveying etc.

CHAPTER 21

FOLDS

The bending of rock-strata due to compressional forces acting tangentially or horizontally towards a common point or plane from opposite directions is known as folding. It results in the crumbling of strata, forming wavy undulations on the surface of the earth, which are known as folds.

Description. The followings are the parts of a fold :

(*i*) The wavy undulations are formed of a series of alternate *crests* and *troughs*.

(*ii*) **Limb.** The stretch of the rock beds lying belween any crest and any of the adjacent troughs on either side is known as the limb of the fold.

(*iii*) **Axial plane.** The imaginary plane which divides the fold as symmetrically as possible is described as its axial plane. Any point on the axial plane is at equidistance from both the limbs. Depending upon the nature of the fold, its axial plane may be vertical, inclined or horizontal or it may even be a curved surface.

(*iv*) **Axis.** In any fold, the line of intersection of the axial plane with the upper and lower surface of any of the constituent beds is known as its axis.

(*v*) **Hinge.** The line along which a change in the amount and/ or direction of dip takes place is known as the Hinge-line and on many folds it coincides with the position of maximum curvature.

The area adjacent to the hinge line is known as the *Hinge-area* or *nose* of the fold.

(*vi*) **Crestal plane.** There is a separate crest for each bed. The plane or surface formed by all the crests is called the crestal plane.

(*vii*) **Trough plane.** The trough is the line occupying the lowest part of the fold. The plane containing such lines may be called the *trough plane*.

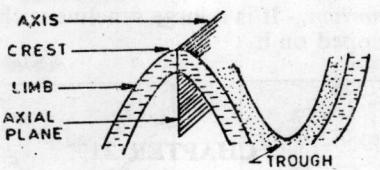

Fig. 21·1

Classification. Folds have been classified into various type, on the following basis :

(a) Appearance in cross-section,

(b) Symmetry of fold,

(c) Thickness of limb,

(d) Interlimb-angle,

(e) Attitude of the fold,

(f) Mechanism of folding,

(g) Origin :   (i) Tectonic-origin,   (ii) Non-tectonic-origin,

(h) Special types.

(a) **Appearance in cross-section**. The following types of folds have been recognised, on this basis.

(i) *Antiform*. Any upwardly convex structure is termed as an antiform. Here the age relationship between the upper and lower set of beds is of complex type.

(ii) *Synform*. Any upwardly concave structure, *i.e.*, flexure in the form of a trough is known as synform.

(iii) *Anticline*. It is generally convex upwards where the limbs commonly slope away from the axial plane. In case of anticlines older beds occur towards the centre of curvature of the fold.

(iv) *Syncline*. It is a fold which is concave upwards and the commonly dip towards the axial plane. Progressive younger beds are found towards the centre of curvature of the fold.

Fig. 21·2

(v) *Anticlinorium*. A large anticline with secondary folds of smaller size developed on it.

(vi) *Synclinorium.* It is a large syncline with secondary folds of smaller size developed on it.

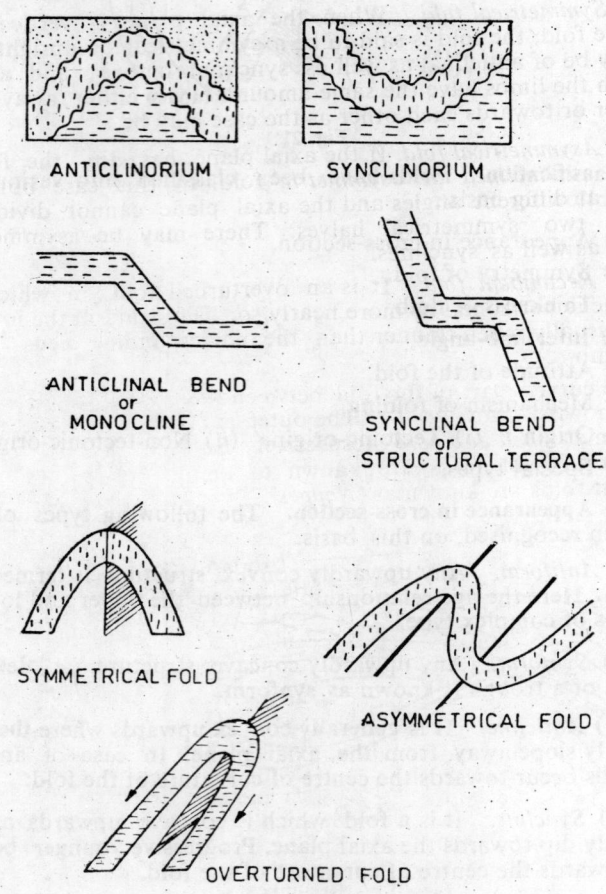

ANTICLINORIUM   SYNCLINORIUM

ANTICLINAL BEND
or
MONOCLINE

SYNCLINAL BEND
or
STRUCTURAL TERRACE

SYMMETRICAL FOLD

ASYMMETRICAL FOLD

OVERTURNED FOLD

Fig. 21·3

(vii) *Anticlinal bend or Monocline.* It is due to local steepening of a bed, whereby there occurs a sudden increase in the dip of a bed, which is originally horizontal, to a near vertical position. But the original bedding remains as before.

(viii) *Synclinal bend or structural terrace.* In case of a dipping bed, due to local flattening of the beds at a particular spot, the beds acquire horizontality and then again follow their original dip without any change in the direction of dip. These are also known as 'structural bench'.

(*b*) **Symmetry of fold.** Six types of folds have been recognised on the basis of symmetry of fold, as follows :

(*i*) *Symmetrical fold.* When the axial plane is vertical and bisects the fold, the fold is said to be a symmetrical or upright fold. They may be of anticlinal as well as synclinal nature ; and accordingly both the limbs have the same amount of dip either away from each other or towards each other as the case may be.

(*ii*) *Asymmetrical fold.* If the axial plane has dip, the fold is described as '*inclined* or *asymmetric*' fold. In this case both the limbs dip at different angles and the axial plane cannot divide the fold into two symmetrical halves. There may be asymmetrical anticlines as well as synclines.

(*iii*) *Recumbent fold* It is an overturned fold, in which the axial plane is horizontal or more nearly so. The strata in the inverted limb are usually much thinner than the corresponding beds in the normal limb.

The curved part of the fold between the normal and inverted limb is known as '*arch bend*'. The outer and inner parts of the fold are known as '*shell* and *core*', respectively. Subsidiary folds attached to the recumbent folds are known as '*digitations*'. Sometimes the recumbent folds are known as '*Nappes*'.

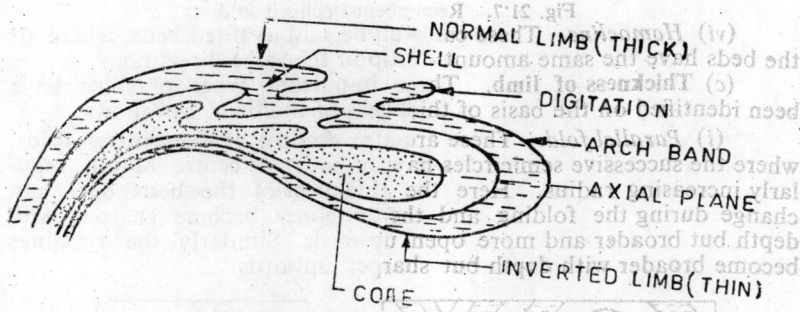

Fig. 21·4. Recumbent fold.

(*iv*) *Isoclinal fold.* In an over-turned fold when both the limbs have the same amount of dip, towards the same direction, it is known as isoclinal fold. There may be vertical, inclined as well as recumbent isoclinal folds.

(*v*) *Over-turned fold.* These are also known as '*over folds*'. Here the axial plane is inclined and both the limbs dip in the same direction, usually at different angles. In this case, one of the limbs occurs in the normal position while the other appears to have been rotated and completely over-turned from its usual position. There may be overturned anticlines as well as synclines.

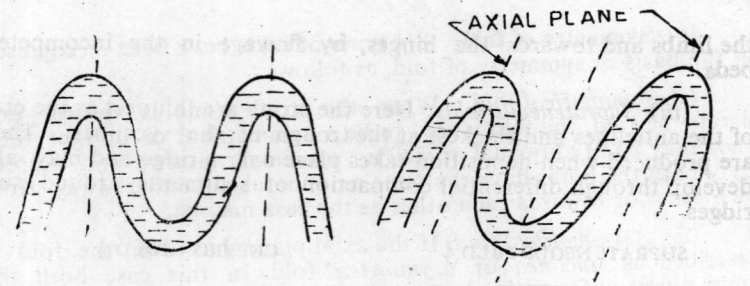

Fig. 21·5. Vertical isoclinal fold.   Fig. 21·6. Over-turned fold.

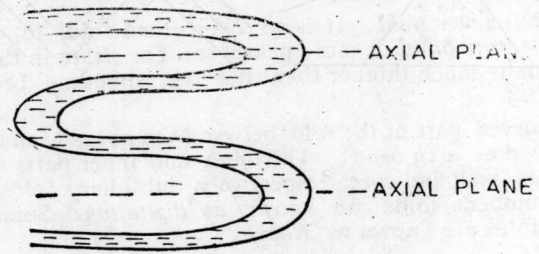

Fig. 21·7. Recumbent-isoclined fold.

(*vi*) *Homocline.* These can well be said as tilted beds, where all the beds have the same amount of dip in the same direction.

(*c*) **Thickness of limb.** Three important types of folds have been identified on the basis of thickness of the limb, as follows:

(*i*) *Parallel fold.* These are also known as concentric folds, where the successive semicircles have a constant centre and a regularly increasing radius. Here the thickness of the beds does not change during the folding and the anticlines become sharper with depth but broader and more open upward. Similarly the synclines become broader with depth but sharper upwards.

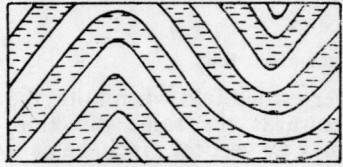
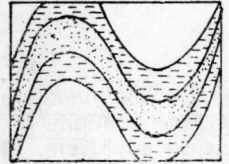

Fig. 21·8. Parallel fold.   Fig. 21·9. Similar fold.

(*ii*) *Similar folds.* In this case the shape of the folds may vary along the axial plane and at right angles to the fold axis. Here every bed is thinner in the limbs and thicker near the hinges. For this, there must be considerable plastic movement of material away from the limbs and towards the hinges, by flowage in the incompetent beds.

(*iii*) *Suprateneous folds.* Here the strata are thinnest at the crest of the anticlines and thickest at the trough of the synclines. These are produced when deposition takes place over a ridge and may also develop through differential compaction of sediments around such ridges.

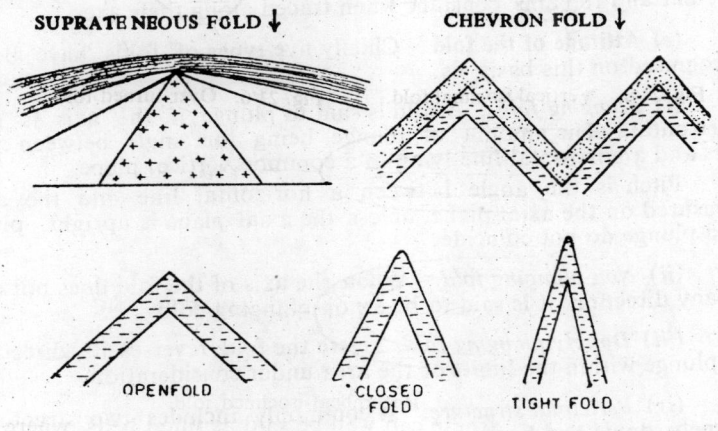

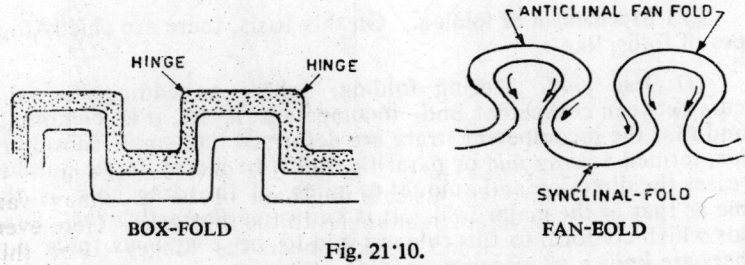

Fig. 21'10.

(*d*) *Inter-limb angle.* On this basis folds have been classified into the following types :

(*i*) *Open or gentle fold.* Where the interlimb-angle is greater than 70°

(*ii*) *Closed fold.* Where the inter-limb-angle is between 30° to 70°.

(*iii*) *Tight fold.* Where the inter-limb-angle is below 30°.

(*iv*) *Cylindrical fold.* When the profile is essentially semi-circular and remains constant when traced along their axes.

(*e*) **Attitude of the fold.** Chiefly five types of folds have been recognised on this basis, as

(*i*) *Plunging fold.* A fold is said to plunge if the axis is not horizontal. The amount of plunge being the angle between the axis and a horizontal line lying in a common vertical plane.

Pitch is the angle between a horizontal line and the axis measured on the axial plane, unless the axial plane is upright, pitch and plunge do not coincide.

(*ii*) *Non-plunging fold.* When the axis of the fold does not dip in any direction, it is said to be a non-plunging fold.

(*iii*) *Doubly-plunging fold.* Here the fold reverses its direction of plunge within the limits of the area under consideration.

(*iv*) *Periclinal structure.* It commonly includes two structures namely, dome and basin.

Dome is an anticlinal structure which plunges in all directions, *i.e.*, is having *quaquaversal* dip. According to some authors, a dome is known as a pericline.

Basin is a synclinal depression with centroversal dip, *i.e.*, dip from all directions towards a contral region. It also known as a *centricline.*

(*v*) *Reclined fold.* Here the axis plunges directly down the dip of the axial plane.

(*f*) **Mechanism of folding.** On this basis, there are chiefly four types of folds, like :

(*i*) *Drag fold.* During folding, where a bedding-plane-slip occurs between competent and incompetent layers, it is commonly found that the incompetent strata are deformed into small subsidiary folds, termed as *drag fold* or parasitic fold. In a very large number of cases the direction and amount of pitch of the drag fold is the same as that of the major fold. It is known as *Pumpelly's rule.* Drag folds which conform to this rule are said to be *congruous* folds and others are known as *incongruous* folds.

(*ii*) *Flexure fold.* It is mostly referred as '*true-folding*'. As it is well known that when compressive force acts on a series of flat beds, there occurs bending, and the convex side is subjected to tension and the concave side is subjected to compression. There is an immediate surface of no strain. Thus the convex side will lengthen, whereas the concave side will shorten and thicken. It involves other developments as

(1) Bending or buckling of the more competent layers under compressive force.

(2) The more passive behaviour of the incompetent beds.

(3) The sliding of beds past one another.

(4) The folding will be analogous to the bending of a thick package of paper.

(*iii*) *Shear-folding.* It is also known as slip folding or 'Gleit-bretter' folding. It results from minute displacement along closely spaced fractures. Each fracture is actually a small fault and the blocks on either side of it have moved upwards progessively.

(*iv*) *Flow-folding.* It develops in highly incompetent beds which behave more as a viscous fluid than a solid rock. These are highly disharmonic in nature. Ptygmatic folding is of this type.

(*g*) **On the basis of origin.** Folds may be of tectonic as well as non-tectonic in origin.

(*i*) *Folds caused by orogenic movements.* If an area has been subjected to more than one phases of folding, there results cross-folding, in which the axial planes and axes of the folds intersect one another at an angle.

If two anticlines belonging to two different sets of folds coincide, the result is a *culmination* but where an anticline and syncline coincides, it results in a *depression*.

Sometimes, a fold is found to increase in size as it is traced upwards along the axial plane ; such a fold is termed as *generative fold.*

(*ii*) *Folds of non-tectonic Origin.*

(1) **Cambering and valley bulging.** Cambering occurs where competent beds form the capping of hills, overlying incompetent beds. The incompetent beds flow outwards into the valley causing the competent beds to downwarp.

In case of valley bulging the incompetent material is forced up into a valley by the weight of the hill masses on either side, which becomes 'turned up' at the edges in the process.

(2) **Diapiric or piercement folds.** These are formed during the upward movement of the mass of rock forming a diapir. Salt domes are the best examples of this type.

(*h*) **Special types of folds :**

(*i*) *Chevron folds.* These are angular folds having straight limbs and sharp hinges. These are also known as *zig-zag, concertina* or *accordion fold* [Fig. 21·10].

(*ii*) *Fan folding.* If, in any fold, both the limbs are overturned the fold assumes the shape of a fan where the crests and troughs are sufficiently rounded. In anticlinal fan folds both the limbs dip towards each other and in synclinal fan folds, away from each other.

(*iii*) *Box fold.* These are rectangular in cross-section. In this crest is broad and flat, two hinges are present, one on either side of the flat crest. These are also known as 'coffer folds' [Fig. 21·10].

(*iv*) *Kink-bands.* These are narrow bands, usually, a few inches or few feet wide, in which the beds assume a dip which is steeper or gentler than the adjacent beds.

(*v*) *Ge'anticline.* It is the broad uplifted area bordering a geosyncline, which supplies sediments for its filling.

(*vi*) *Geosyncline.* These are elongated large basins which are found submerged beneath the sea-water and contain very great thickness of sediments.

Recognition in the field :

(*i*) Folds are recognised in the field, by plotting the attitude of the beds *i e.*, their dip and strike on the map.

(*ii*) When the types of folds are to be recognised on the basis of the age consideration of the beds, the top and bottom of beds are determined by taking into account the nature, and form of the features occurring on the beds itself. Accordingly some types of folds have been inferred.

(*iii*) In railway cuttings, excavation of tunnels as well as drilling of rock beds for certain purposes clearly reveal the occurrence of folded-structures in the field.

Palinspastic maps place rocks in their presumed position before folding and thrusting.

Effects on outcrops :

(*i*) Folds cause shortening of the crust of the earth and their subsequent thickening.

(*ii*) It is usually observed that streams follows the axis portion of the anticlinal ridges and high-lands and domed-structures occur along the axis of the synclines.

(*iii*) Repetition of beds in their occurrence in the field, infers the presence of a fold.

CHAPTER 22

FAULTS

Faults are well-defined cracks along which the rock-masses on either side have relative displacement. The attitude of faults are defined in terms of their strike and dip. The strike and dip of a fault are measured in the same way as they are for bedding.

Description. The followings are the parts of a fault :

(*i*) **Footwall and hanging-wall.** Of the two blocks lying on either side of the fault-plane, one appears to rest on the other. The former is known as hanging-wall side while the latter which supports the hanging-wall is known as the footwall side.

(*ii*) **Fault scrap.** The relative displacement on either side of the fault line results in an upstanding structure with a steep side which is called *'fault scrap'*.

Fault-line scrap. It owes its relief due to differential erosion along a fault-line.

(*iii*) **Down thrown side and up-thrown side.** In case of a fault, one of the dislocated block appears to have been shifted downwards in comparison with the adjoining block lying on the other side of the fault-plane. The former, therefore is known as the down-thrown side while the latter is described as the up-thrown side.

Terminologies associated with faults :

1. **Strike.** Strike of the fault is the trend of a horizontal line in the plane of the fault.

2. **Dip.** Dip is the angle between a horizontal surface and the plane of the fault and is measured in a vertical plane that strikes at right angles to the fault.

3. **Hade.** It is the complement angle of dip, *i.e.*, the angle which the fault plane makes with the vertical plane or ($90° - $Dip $=$Hade).

4. **Throw and heave.** The throw of a fault is the vertical component of the apparent displacement of a bed, measured along the direction of dip of the fault.

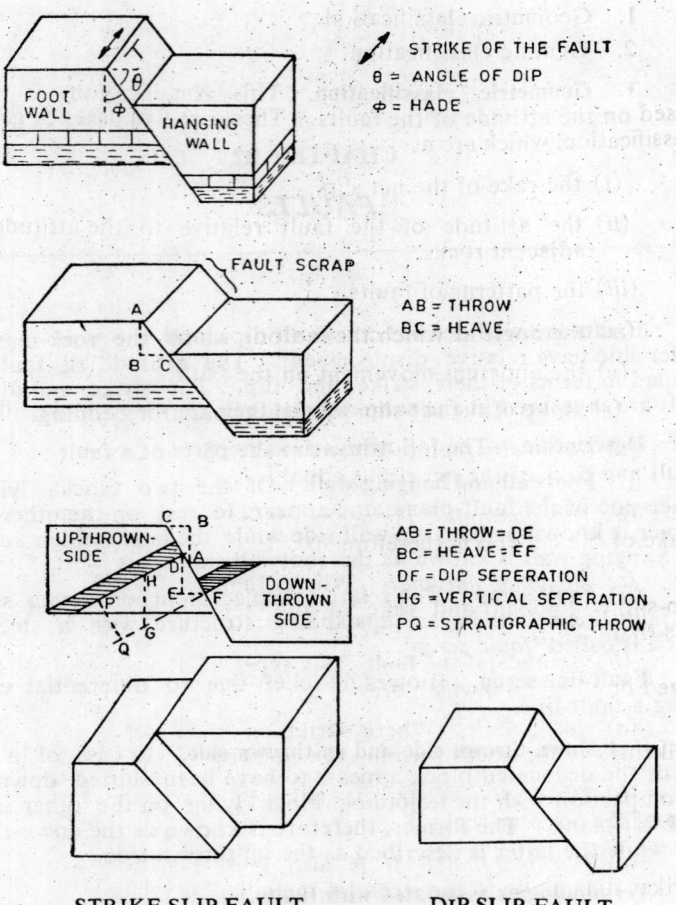

STRIKE OF THE FAULT
θ = ANGLE OF DIP
φ = HADE

FOOT WALL

HANGING WALL

FAULT SCRAP

AB = THROW
BC = HEAVE

UP-THROWN SIDE

DOWN-THROWN SIDE

AB = THROW = DE
BC = HEAVE = EF
DF = DIP SEPERATION
HG = VERTICAL SEPERATION
PQ = STRATIGRAPHIC THROW

STRIKE-SLIP FAULT   DIP SLIP-FAULT

Fig. 22.1

The heave of a fault, in a like manner, is the horizontal component of the apparent displacement. It is also known as *gape*.

5. Stratigraphic throw. If the same bed occurs twice because of faulting, the perpendicular distance between them measured along a vertical section at right angles to the strike of the fault, is known as stratigraphic throw.

6. Net-slip. The total displacement due to a fault is described as its *net-slip*.

Classification of faults. There are two type of classification of faults :

1. Geometric classification.
2. Genetic classification.

1. **Geometric classification.** This classification is strictly based on the attitude of the faults. There are five bases of geometric classification, which are as

(*i*) the rake of the net slip,

(*ii*) the attitude of the fault relative to the attitude of the adjacent rocks,

(*iii*) the patterns of faults,

(*iv*) the angle at which the fault dips, and

(*v*) the apparent movement on the fault.

(*i*) **Rake of the net-slip.** On this basis folds are classified as

(*a*) *Strike slip fault* Where the net slip is parallel to the fault and rake of the net slip is equal to zero.

(*b*) *Dip-slip fault.* Here the net slip is equal to the dip-slip. Rake of the net-slip is therefore 90°.

(*c*) *Diagonal-slip fault* Where there is both a strike-slip and dip-slip component and rake of the net slip is more than '0°' but less than '90°'.

(*ii*) **Attitude of the fault.** Six-types of faults have been recognized on this basis, which are

(*a*) *Strike-fault.* Where strike of the fault is parallel to the strike of the rock-beds forming the country.

(*b*) *Dip-fault.* Where the strike of the fault is parallel to the dip of the country rocks.

(*c*) *Diagonal fault.* It is also known as oblique fault, which strikes diagonally to the strike of the adjacent rocks.

(*d*) *Bedding fault.* In this case the fault plane is parallel to the bedding planes of the adjacent rocks.

(*e*) *Longitudinal fault.* Here the fault strikes parallel to the strike of the regional structure.

(*f*) *Transverse fault.* It strikes perpendicularly or diagonally to the strike of the regional structure.

(*iii*) **Fault-pattern.** On this basis the following types of faults have been recognised :

(*a*) *Parallel faults.* It consists of a series of faults having the same dip and strike.

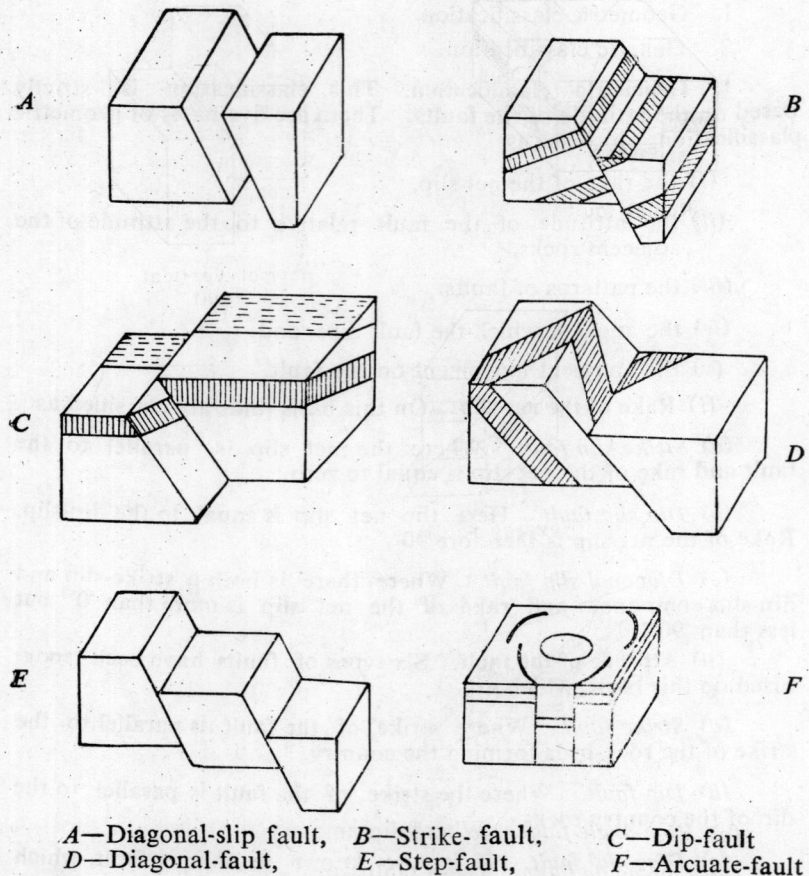

A—Diagonal-slip fault,   *B*—Strike- fault,   *C*—Dip-fault
D—Diagonal-fault,   *E*—Step-fault,   *F*—Arcuate-fault
Fig. 22·2

(*b*) *Step-faults*. If in a series of parallel faults the successive blocks are down-thrown more and more towards a particular direction, the resulting structure will be a step-fault.

(*c*) *Arcuate fault*. These are also known as peripheral faults which have circular or arc-like out-crop on a level surface.

(*d*) *Radial faults*. Here a number of faults belonging to the same system, radiate out from a point.

(*e*) *Enechelon faults*. There are relatively short faults which overlap each other.

(*iv*) **On the basis of dip value.** Two important types of faults have been recognised on this basis. They are as

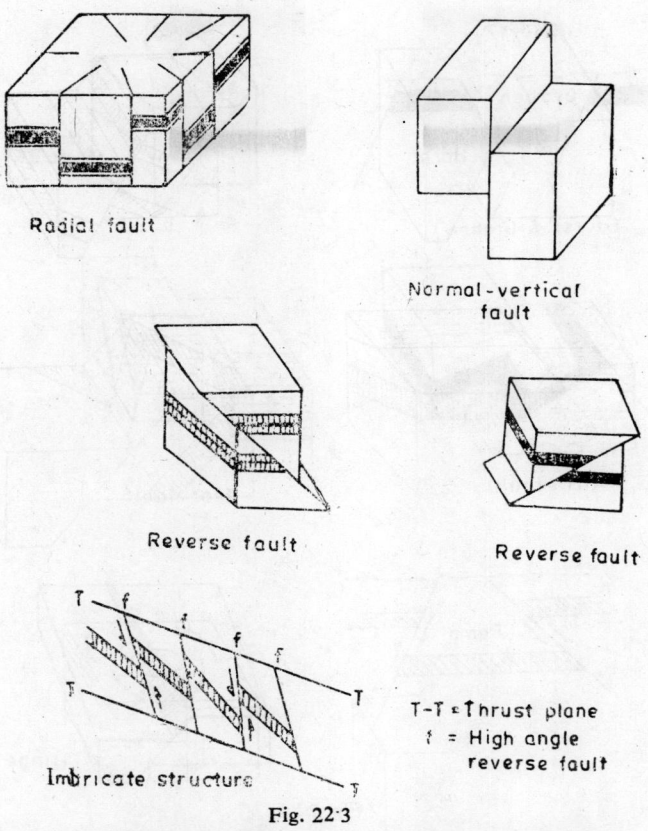

Radial fault

Normal-vertical fault

Reverse fault

Reverse fault

T-T = Thrust plane
f = High angle reverse fault

Imbricate structure

Fig. 22·3

(a) *High-angle fault.* Where dip amount is more than 45°.

(b) *Low-angle fault.* These faults dip less than 45°.

(v) **Apparent-movement.** On this basis faults can be classified into :

(a) *Normal faults.* Which are inclined faults in which the hanging-wall side appears to have moved relatively downwards in comparison to the adjoining foot-wall side.

(b) *Reverse faults.* In this case the foot-wall side appears to have been shifted downwards in comparison to the hanging walls.

Besides the above types, there is an important type of fault, known as the *pivot* or *scissor fault* or *hinge-fault*. In this case one block appears to have rotated about a point on the fault plane such that for part of its length the fault, is normal with a decreasing throw

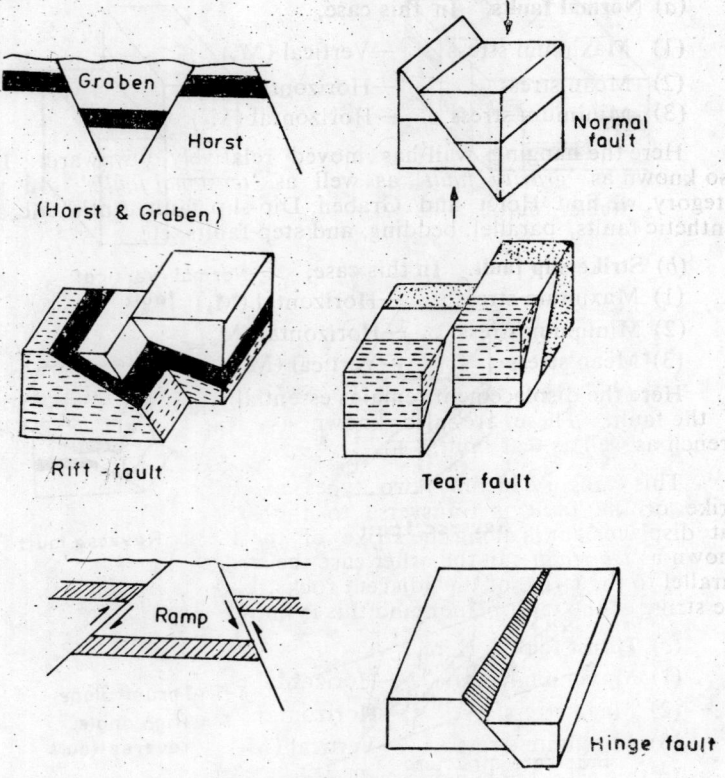

Fig. 22·4

and for the remainder of its length is a reverse fault with an increasing throw. The position of no displacement being the point around which rotation appears to have taken place.

2. Genetic classification. It is well known that along the shear fractures the displacement is parallel to the walls and there is no movement perpendicular to the fracture. It is assumed that the displacement are caused by some stresses. Three types of principal stresses have been assumed. Of the three principal stresses two are horizontal and the third one is vertical and due to gravity alone. Three sets of conditions in which all the stresses are compressional may arise ; accordingly three sets of faults originate. Besides, this classification considers whether the forces are compressional, tensional, shearing or torsional.

Three main types of faults have been recognized basing on the orientation of the three principal stresses.

(*a*) **Normal faults.** In this case,

(1) Maximum stress   —Vertical (M_1)

(2) Mean stress   —Horizontal (M_2)

(3) Minimum stress   —Horizontal (M_3)

Here the hanging wall has moved relatively downward. It is also known as '*gravity fault*' as well as '*tensional faults*'. In this category, we find Horst and Graben, Dip-slip fault, Antithetic and Synthetic faults, parallel, bedding, and step-faults, etc.

(*b*) **Strike-slip fault.** In this case,

(1) Maximum stress   —Horizontal (M_1)

(2) Minimum stress   —Horizontal (M_3)

(3) Mean stress   —Vertical (M_2)

Here the displacement remains essentially parallel to the strike of the fault. These are also known as transcurrent, transform, wrench as well as tear faults.

This category includes two types of faults; in one case, the strike of the fault is transverse to the strike of the country rock but displacement is along the strike of the fault plane, where it is known as *tear fault*; in the other case the strike of the fault plane is parallel to the strike of the adjacent rocks (but the net slip is along the strike of the fault plane), and this is known as rift-fault.

(*c*) **Thrust fault.** Here,

(1) Maximum stress   —Horizontal (M_1)

(2) Mean stress   —Horizontal (M_2)

(3) Minimum stress   —Vertical (M_3)

In this case, the hanging wall moves relatively over the foot-wall. It includes the following types of faults:

(*a*) **Reverse fault.** Where the dip of the fault is more than 45°.

(*b*) **Thrust faul** Where the fault has a dip less than 45°.

Thrusts are again subdivided into two types:

(*i*) *Over thrust.* In which the initial dip is 10° or less and the net-slip is measured in terms of miles.

(*ii*) *Under thrust.* In this case, the foot-wall side actually moved and pushed itself underneath the hanging-wall side.

(*c*) Nappes.

(*d*) Imbricate or Schuppen structures, etc.

Schuppen structure. When several thrust planes develop in parallel sets, a series of high angle reverse-faults may also develop between pairs of thrust planes, giving rise to what is known as 'imbricate or schuppen structure'.

Recognition of faults in the field and their effects on outcrops. To recognise faults in the field, a number of criteria are used. The faults may be directly seen in the field, particularly in artificial exposures such as river-cuttings, road cuttings, etc. But in majority of cases, faults are recognised by *stratigraphic* and *physiographic* evidences as

1. Discontinuity of structures.
2. Repetition or omission of strata.
3. Silicification and mineralisation.
4. Features characteristic of fault-planes.
5. Sudden change in the sedimentary facies.
6. Physiographic evidences, etc.

1. Discontinuity of structure. If there *is* an abrupt termination of any geologic structure, a fault can be expected near about the point of termination. Sudden termination of dykes and veins, etc. also suggests the existence of a fault.

2. Repetition and ommision strata. Sometimes a bed may suddenly terminate but occurs again somewhat off from the place where it terminated besides sometimes rock beds forming the country are found to be repeated and/or omitted (but maintaining the same order) indicates the occurrence of a fault.

3. Silicification and mineralisation. Faults are often the avenues for moving solutions. The solution may replace the country rock with fine-grained quartz, causing silicification and sometimes also they form mineral deposits at that site. It points to the occurrence of a fault in that area.

4. Features characteristic of fault plane. These features are produced due to friction of the blocks on either side of the fault plane and includes features like slickensides, grooves, drag or fluccan (*i.e.*, pulverized clayey matter), breccia (commonly known as crush breccia), mylonites, horses and slices etc.

5. Difference in sedimentary facies. Different sedimentary facies of rocks of the same age may be brought into juxtaposition by large horizontal displacements, is suggestive of faulting.

6. Physiographic evidences. The 'effects of faulting on outcrops' constitute the physiographic evidences for faulting. They include features like :

(*i*) Offset ridges, (*ii*) Fault scrap, (*iii*) Piedmont scrap, (*iv*) Tringular facets, (*v*) Offset streams, (*vi*) Springs following a straight course (sometimes springs aligned along the fault planes also), (*vii*) Lineament suggest the presence of faults in the field (*viii*) Alluvial fan, (*ix*) Monocline, etc.

CHAPTER 23

JOINTS

A joint is defined as a fracture in a rock between the sides of which there is no observable relative movement. They are present in most consolidated rocks of igneous, metamorphic and sedimentary origin. Joints may form as a result of either diastrophism or contraction.

Description :

(i) A series of parallel joints is called a '*joint set*'.

(ii) Two or more joint sets intersecting each other produce a '*joint system*'.

(iii) Two sets of joints nearly at right angles to one another, produced by the same stress system, is known as conjugate system.

(iv) A persistent joint or set which may be horizontal or vertical, is called '*master joint*'.

Classification :

1. According to the mode of origin, three types of joints have been recognised, as follows :

(a) **Tensional joints**. These are also known as '*shrinkage joints*'. In igneous rocks, they are produced as a consequence of contraction due to cooling. '*Columnar Structure*' which characterises many basic extrusive and intrusives, consists of long hexagonal blocks closely packed together.

In granites and granodiorites several sets of joints may be observed, but commonly three sets are prominent—one horizontal and two vertical at right angles to each other and to the horizontal set. If these sets are more or less equally spaced, the fracture planes give rise to cubical blocks ; the jointing is then termed '*Mural jointing*'.

Joints formed in little deformed sedimentary rocks are due to tension caused by compaction and shrinkage as sediments are consolidated into sedimentary rocks. Tensional joints may also be due to deformation.

(b) Sheet joints. These joints develop in sets and are more or less parallel to the surface of the ground, especially in plutonic igneous intrusions such as granite. They may originate due to unloading of the rock mass when the cover is removed through the processes of erosion.

(c) Tectonic joints. These are also known as *shear joints*. They are formed in a rock under compression. They originate as a direct result of folding or thrusting in rocks. Generally they are of three types :

(i) *Strike set.* Longitudinal joints parallel to the fold axis.

(ii) *Dip-set.* Also known as cross-joints, perpendicular to the longitudinal joints.

(iii) *Diagonal set.* Which is a conjugate set of oblique joints, which lie at rather less than 45° to the direction of tectonic-axis.

2. According to the geometric classification of joints, there are three important varieties, like strike-joints, dip joints and diagonal joints, which is totally with respect to the regional strike and dip of the country rocks.

Joints may be open or closed. The closed joints are also known as latent, blind or incipient joints. They may become open as a result of weathering, which is commonly found in jointed limestones.

Recognition of joints in the field and their effects on out crops. Joints are generally recognised in the field as 'faults without displacement'. Their dimension varies within wide-limits. Sometimes they are very short in their extension, but in certain cases they are found to extend for miles to-gether.

Joints commonly control the drainage pattern of an area. They also determine the shape of coastlines, because they provide a passage, whereby water may penetrate deeply into the rockmass, thus allowing weathering to take place.

Jointed rocks are pervious to fluids and hence may act as aquifers or reservoir rocks for oil or natural gas.

The presence or absence of joints in a region matters much to quarrymen and miners because it determines the ease with which querry and mining can be accompolished.

Sometimes joints act as avenues for molten rock materials to come above the surface. It also determines the localisation of some mineral deposits.

CHAPTER 24

UNCONFORMITIES

Introduction. An unconformity is a plane of discontinuity that separates two rocks, which differ notably in age. The younger of these rocks are nearly always of sedimentary origin and must have been deposited on the surface of the older rock, which is a surface of erosion.

How it forms. The formation of an unconformity may be attributed to three main processes like erosion, deposition and tectonic-activity. The stages of its development involves.

(*i*) The formation of older rocks.

(*ii*) Upliftment and subaerial erosion of the older rocks.

(*iii*) The formation of a younger succession of beds above the surface of erosion.

Thus an unconformity is a surface of erosion or non-deposition that separates younger strata from older ones. It is accordingy regarded as a planar structure.

Description. Usually the following characteristics are found to be associated with the unconformities :

(*a*) There is a difference in the lithological composition, thickness and order of superposition of the overlying strata with respect to underlying rocks.

(*b*) Difference in age as indicated by the fossil assemblages of the overlying and underlying beds.

(*c*) Sometimes the underlying beds possess different dip and strike value than that of the overlying beds.

(*d*) The presence, in most cases, of a conglomerate horizon at the bottom of the younger set of beds.

Classification. Unconformities have been classified into various types on the basis of the factors like :

(*i*) The relationship existing between the underlying and abovelying rock beds. *i.e.*, whether both are sedimentary or one of them is of igneous origin.

(*ii*) The attitude of the underlying and overlying beds.

The various types of unconformities may be enumerated as follows :

    (*a*) Angular Unconformity.

    (*b*) Disconformity.

    (*c*) Local Unconformity.

    (*d*) Non-conformity.

    (*e*) Blended Unconformity.

ANGULAR CONFORMITY

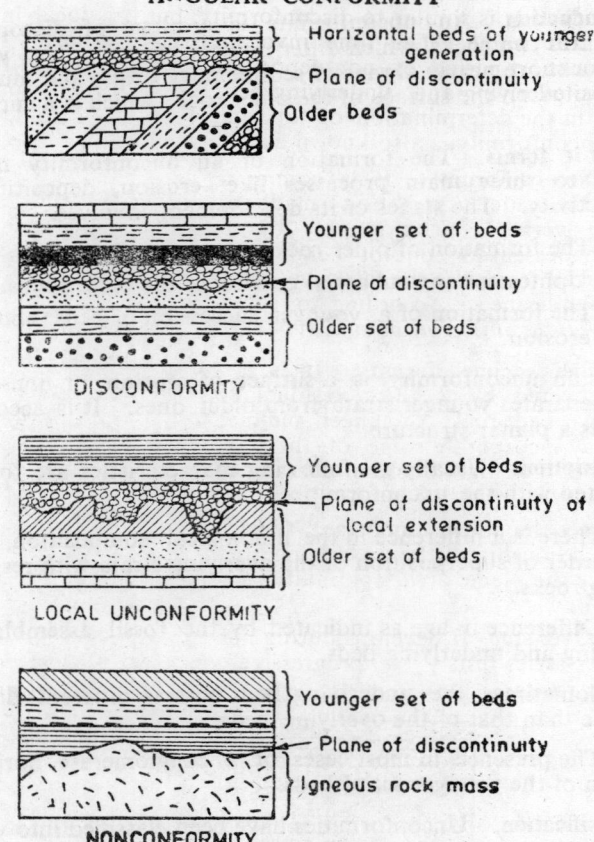

Fig. 24·1

(*a*) **Angular unconformity.** If the beds beneath the erosion surface are folded or tilted so that there is an angular discordance between the younger and older beds, the contact is called an angular uncon-formity. In this case, both the underlying and overlying rocks are of sedimentry origin, but the attitude of the rocks above and below the plane of discontinuity differs from each other.

(*b*) **Disconformity.** It is also known as '*Parallel-unconformity*' in view of the fact that the bedding above and below the plane of discontinuity. The lower and upper series of beds dip at the same amount and in the same direction, thus this type or unconformity is formed when there is a lesser magnitude of diastrophism, or distur-bance between the deposition of the two succession of strata.

(*c*) **Local-unconformity** It is also known as a '*non-depositional unconformity*'. It is similar to disconformity, but it is local in extent and hence the name. The time involved is also short. Thus it represents a short period of non-deposition. So the age difference between the overlying and underlying beds is very less. This is quite significant in the determination of top and bottom of folded-beds. Such an unconformity is also known as '*Diastem*'.

(*d*) **Non-conformity** It is commonly appiled to structures in which the older formation made up essentially of plutonic rocks, is overlain unconformably by sedimentary rocks or lava-flows. The essential concept being that prolonged erosion must have occurred to expose the intrusive before burial. It is not of any tectonic significance. According to some geologists, it should be termed as '*Heterolithic unconformity*.'

(*e*) **Blended unconformity.** It is a surface of erosion, which may be covered by a thick residual soil that grades into the underlying bed rock. Younger sediments deposited above the surface may incorporate some of the residual-soil and a sharp con-tact may be lacking. Such a contact may be called as Blended-unconformity.

Besides the above important types of unconformity, various types of relationship of the underlying beds with that of the beds lying above the plane of discontinuity has been recognised, which are as follows :

1.   **Over-step.** It develops during marine transgression, so that the younger series rests   progressively on older members of the underlying rocks.

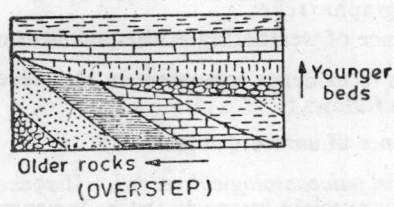

Older rocks
(OVERSTEP)

Younger
beds

Fig. 24·2

2. Over-lap. It is often found that younger bed completely covers up and advances much beyond the limits of the underlying rocks. It is best observed along the gently sloping fringes of inland basins during marine transgression.

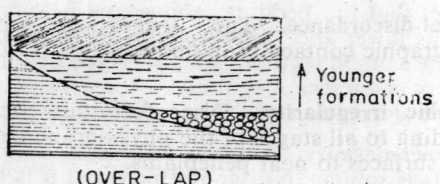

(OVER-LAP)

Fig. 24·3

3. Off-lap. It is reverse of overlap, *i.e.*, lower beds of the upper series extends further than the younger ones due to marine regression.

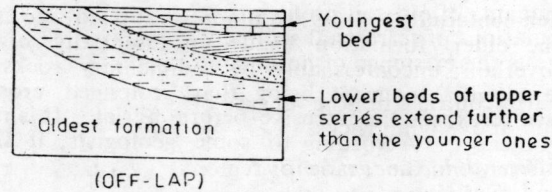

(OFF-LAP)

Fig. 24·4

4. On-lap. It appears synonymous with overlap or sometimes overlap + overstep.

Recognition of unconformity in the field and their effects on outcrops. Unconformities may be recognised in various ways, of which observation in a single outcrop is the most satisfactory. Reliability in the identification of an unconformable relation, in general, increases in proportion to :

(*a*) Time interval.

(*b*) Thickness of beds missing from the stratigraphic record.

(*c*) Structural discordance.

(*d*) Topographic relief.

(*e*) Evidence of weathering at the unconformable surface.

The kinds of evidence important in the recognition of unconformities are as follows :

1. Evidence of unrecorded interval :

(*i*) *Gap in palaeontological record.* If rocks with upper triassic fossils is directly overlain by rocks with lower cretaceous fossils, it is said to be an uncoformity.

(ii) *Gap in stratigraphic record.* Good evidence of unconformity may be provided by the local absence of distinctive strata characterised by either lithologic or palaeontologic peculiarities or both. Abrupt changes in these characters may suggest the presence of an unconformity.

2. Structural-discordance. The truncation of sedimentary layers at a stratigraphic contact is an unmistakable evidence of discontinuity.

3. Topographic irregularity. Disconformities preserve topography corresponding to all stages of the physiographic cycle ranging from undissected surfaces to near peneplains.

4. Evidence of old-land surface :

(1) Weathering may aid in the recognition of old land surface and the unconformites that mark their position in the stratigraphic section.

(2) Remnants of old-soil profile are more likely to be preserved in the accumulations of terrestrial sediments. Their presence within a strata suggests the existence of an unconformity.

5. Sharp contrast in the degree of induration between the underlying and above lying rocks.

6. Difference in the grade of regional metamorphism of the underlying and overlying rocks.

7. Significant difference in the intensity of folding in both younger and older set of beds.

8. Presence of basal conglomerate horizon suggests the existence of an unconformity.

In the outcrop, unconformities are usually found as a rugged and eroded surface.

CHAPTER 25

DETERMINATION OF THE ORDER OF SUPERPOSITION IN THE FIELD

As we know in any sedimentary sequence, the rock beds exist in order, from the bottom to the top of the column. The bottom most bed is naturally the oldest one while the top most is the youngest. But where the sedimentary columns have undergone very severe diastrophism, it is difficult to determine the order of superposition. The following criteria may be used for determining the order of superposition in the areas, which have been tectonically deformed.

Primary Sedimentary Structures :

(*a*) **Ripple marks.** Only oscillation or symmetrical ripple marks which have rounded troughs and pointed crests are mostly used for the said purpose. In case of inverted sequence the tapering crest points downward and rounded troughs point upwards.

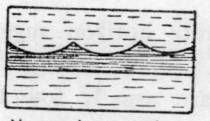
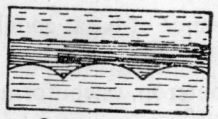

Normal condition   Over-turned

A   B

Fig. 25·1

(*b*) **Current bedding.** Since these beds are usually found to lie

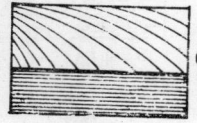

Normal condition   Over-turned-position

A   B

Fig. 25·2

slightly oblique to the major bedding planes, these are conveniently used for determining the top and bottom of beds.

(c) **Graded bedding.** As it shows a gradation in the size of grains from coarse below to fine above, is used as an important criteria.

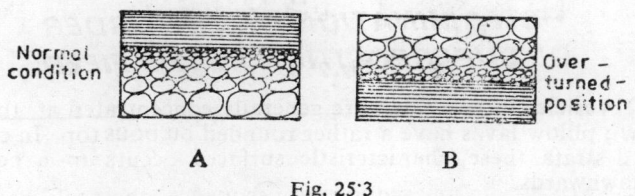

Fig. 25·3

(d) **Mud-cracks.** These are usually wide at the top and taper downwards but in inverted sequence the tapering end faces the upward direction.

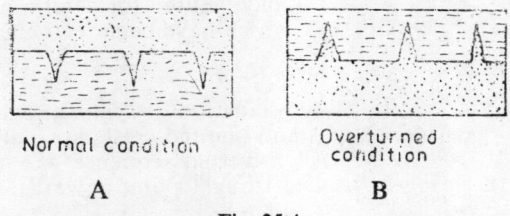

Fig. 25·4

(e) **Rain-prints, tracks and trails and foot prints** of organisms all form depressions on the upper surface of the bed and the concave portion faces upwards whereas in the inverted strata the concave portion faces downward.

(f) **Local unconformities ;** also gives conclusive evidence about the order of superposition of strata. Similarly channels cut into the underlying beds also indicates the top and bottom of original beds.

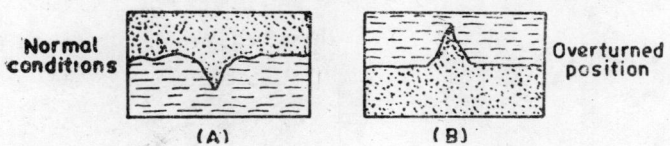

Fig. 25·5

(g) **Orientation of fossil-shells.** Certain types of shells tend to lie on the sea-floor in their most stable orientation, for example lamellibranch shells and 'Productid'-brachiopods lie on bedding planes with their convex surface facing upward. In an inverted sequence, the concave surface faces upward.

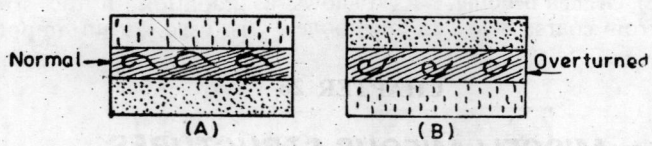

Fig. 25.6

(*h*) Vesicles in lava flows are generally concentrated at the top of a flow ; pillow lavas have a rather rounded bulbous top. In case of deformed strata these characteristic surfaces occurs in a position facing downwards.

CHAPTER 26

MISCELLANEOUS STRUCTURES

(*a*) **Outliers and inliers.** A limited area, where younger rocks are completely surrounded by older rocks, it is known as an *outlier*. In a similar way, an area in which the older rocks are completely surrounded by rocks of younger age, known as '*inliers*'.

Chiefly three processes can be attributed to the formation of both outliers as well as inliers, which are as follows :

1. Folding. Folds which attains the form of anticlines and synclines, after undergoing a period of erosion at the surface produces outliers and inliers.

Anticlinal folds produce inliers and synclinal folds produce outliers.

2. Faulting. Because of the dislocation of rocks due to faulting, sometimes it so happens that older rocks are surrounded by younger ones or the *vice-versa* and accordingly produces outliers or inliers.

3. Erosion. Erosion plays a major role in the formation of outliers and inliers only in cases where the disposition of the beds are horizontal. Sometimes deep valleys are being cut by the agents of erosion and when the erosion exposes underlying rocks around the younger-remnant bed, the resulting structure is termed as *outlier*, this process is equally applicable for the formation of *inliers*.

(*a*) **Nappes and windows.** Nappe is a large body of rock that has moved forward for more than one mile, from its original position either by overthrusting or by recumbent folding. Because of the translation of rocks in nappes, various types have been recognised :

(*i*) *Autochthonous nappes*, where the rock beds have not been translated.

(*ii*) *Para-autochthonous* nappe, in which the sheet of rocks have undergone relatively small translation and which can be traced back to their origin or roots.

(iii) *Allochthonous nappe*, here the rock sheets have been translated for great distances.

Window. Sometimes through erosion of overlying nappes the younger rock beneath it becomes exposed. Thus, the area where erosion has penetrated a nappe (*i.e.*, both large recumbent folds as well as thrust) exposing rock lying beneath, is known as a window or *fenster*.

(c) **Klippe.** With continued erosion of a nappe, and window sometimes a remnant of the thrust sheet is left as a relict block, which is called a 'klippe' (*cliff*).

Geological Surveying and Mapping—Use of Contour and Topogrphical Maps'

Important Points :

(1) **Contour.** These are imaginary lines which joins points of equal elevation. The representation of relief features on a map is being made through drawing of contour lines.

The contour lines when are found circular, it represents a hill or mountain. But when they are circular and at the same time closely spaced, it indicates a vertical or nearly vertical peak. The contour taper in any direction to show a valley or river in that direction.

(2) **Topographical maps.** These maps are to represent the surficial relief features of a particular area (both positive and negative) on a sheet of paper. It includes not only the cartographic representation of hills, mountains and rivers but also other features like temples, location of villages etc.

With the use of clinometer compass, by the back-bearing method, one can locate his position in the field with reference to map of the area.

Geological mapping of an area includes the methods by which exposures of various rock types as recognised in the field and their attitude are represented in the topographical map. Accordingly the structure of the area from the geological point of view and the occurrences of various rock types of a particular can be identified from its geological map.

In case of Geological surveying, the representation of the important features of an area, is being made and it indicates the location of various important objects, along with their direction, in a particular area according to a given scale.

PART IV
CRYSTALLOGRAPHY

CHAPTER 27

ELEMENTS OF CRYSTAL-FORMS AND SYMMETRY

Crystals are solid geometric figures which are bounded by well defined more or less plane surfaces called 'faces'. Crystals show the following general characteristics :

(i) Crystals are polyhedral bodies.

(ii) It possesses a typical internal atomic structure and accordingly the faces of a crystal are arranged in a regular pattern.

(iii) The regular geometry is developed only under suitable physico-chemical condition. Crystals are formed due to slow-cooling.

(iv) A crystal which possesses both external form as well as internal atomic structure is said to be a 'perfect crystal' ; but if it possesses only the internal atomic structure without the development of corresponding external form, it is said to be 'crystalline'; when there is neither internal atomic structure nor external form, it will be known as an *amorphous substance*.

Elements of Crystal Forms :

(a) **Faces.** These are the external expressions of the internal atomic planes of the crystal. Faces are said to be 'like' when they have similar properties and 'unlike' when they show different characteristics.

(b) **Edge.** The line of intersection of two adjacent faces is known as an edge.

(c) **Solid angle.** It is a point where three or more faces meet.

(d) **Form.** It is a group of faces, which have a like position with respect to the crystallographic axes of referrence.

Forms may be classified in three ways :

I. (i) **Simple form.** When a crystal is made up of all like faces such as cube, octahedron, etc.

(*ii*) **Combination form.** When a crystal is made up of two or more simple forms, such as when it consists of basal pinacoid and prism faces, each of which in itself is a simple form.

II. (*i*) **Open form.** These forms whose faces cannot enclose space all by themselves, as they do not have adequate number faces to do so and as a result occur only in combination with other forms, such as pinacoids and prisms.

   (*ii*) **Closed form.** It is an assemblage of faces, which can enclose a volume of space.

III. (*i*) **General form.** It is one in which the indices are unrestricted in magnitude.

   (*ii*) **Special form.** Here only one possibile set of values exist for the indices (*hkl*), *e.g.*, only one octahedron (III) is possible in the cubic.

   (*iii*) **Restricted form.** When the forms are neither special nor general-part of their index is variable and part fixed, *e.g.*, in a prism (*hkl*), '*l*' must always be zero and in trisoctahedra of the cubic system, '*h*' must always equal '*k*', this type of forms are called restricted forms.

Besides the above classification, forms have also been classified as

(*a*) **Holohedral forms.** These forms exhibit the highest degree of symmetry possible in a system.

(*b*) **Hemihedral forms.** These forms show half the number of faces required for the full symmetry of the system, *e.g.*, tetrahedron is a hemihedral form of octahedron.

(*c*) **Hemi-morphic forms.** These forms have dissimilar faces about the two ends of an axis of symmetry. This axis is called the polar axis. Thus half of the faces of a holohedral form are grouped about one end of the axis and none at the other. Hemimorphic forms lack centre of symmetry.

(*d*) **Tetartohedral forms.** They show only a quarter of the number of faces of the corresponding holohedral form. These forms have neither plane nor centre of symmetry.

(*e*) **Enantiomorphic forms.** These forms do not have either plane or centre of symmetry and occur in two positions which are mirror images of each other. They cannot be converted into each other by any rotation whatsoever.

Common Forms in Crystallography :

(*i*) **Pedion.** It is represented by one face only.

(*ii*) **Pinacoid.** It is an open form, consisting of two faces which cuts one crystallographic axis and remains parallel to the remaining axes.

(*iii*) **Prism.** It is also an open form, consisting of four faces, each face of which essentially parallel the vertical axis and cuts one or more horizontal axes.

(*iv*) **Pyramids.** It is a closed form having eight faces, each face of which cuts the vertical axis and cuts one or more horizontal axes, at equal or unequal distances.

(*v*) **Domes.** It is an open form intermediate between a prism and a pyramid, whose faces cut the vertical axis and one of the horizontal axes. These are also known as '*Horizontal prism*'.

(*vi*) **Diametral Prisms.** It is formed by the combination of three pinacoids which together enclose space. They occur only in the Orthorhombic , Monoclinic and Triclinic systems in which all the pinacoids occur.

Elements of Symmetry :

Crystals also show certain regularity of positions of faces, edges, corners, solid-angles etc. The geometric locus about which a group of repeating operations act is known as a symmetry element. Sometimes the repetition is with respect to a point, in which case, it has '*centre of symmetry*', sometimes it is with respect to a line, in which case, it has an '*axis of symmetry*' and when the repetition is with respect to a plane, it is said to have a '*plane of symmetry*'.

1. **Centre of symmetry.** The point within a crystal through which straight lines can be drawn so that on either side and at the same distance from the centre similar faces, edges and solid angles are encountered, is known as the centre of symmetry. In other words, a crystal is said to possess a centre of symmetry, when for each face, edge, corner etc., on one side of the crystal, there is a similar face, edge or corner, directly on the opposite side of the centre point.

2. **Axis of symmetry.** It is an imaginary line about which if the crystal is allowed to rotate through an angle of 360°, similar faces, edges and solid angles will come to the space for more than once. If it comes twice, the axis is an axis of 'two fold' symmetry ; if it occurs thrice, it is an axis of three-fold symmetry.

The maximum number of axis of symmetry is '13' and it is found in Isometric system.

3. **Plane of symmetry.** It is an imaginary plane which passes through the centre of the crystal and divides it into two parts, such that one part is the mirror image of the other. These planes of symmetry may be diagonal, horizontal as well as vertical.

There are maximum nine planes of symmetry, which is the normal class of isometric system.

Symmetry elements have a particular relationship with the internal atomic structure of the crystals. Accordingly, they form the basis for the classification of crystals into thirty-two symmetry classes.

It is quite significant to note that the normal classes of all the systems show maximum number of symmetry elements, but the other classes consisting of hemihedral, hemi-morphic, tetarohedral and enantiomorphic forms show minimum number of symmetry elements in comparison to the normal class which consists of the holohedral forms.

Pseudo-symmetry. Crystals of certain species imitate the symmetry of a class or a system higher or lower than to which they actually belong. It may be due to twinning, distortion or by imitation of interfacial angles.

CHAPTER 28

LAWS OF CRYSTALLOGRAPHY

Through studies of external forms and angular relationships between the crystal faces, some fundamental laws have been established, which govern the whole crystallography. They are as follows :

1. Law of constancy of interfacial angle.
2. Law of rational indices.
3. Law of axial ratio.
4. Law of crystallographic axes.
5. Law of constancy of symmetry.

1. Law of constancy of 'interfacial-angle'. Interfacial angle may more generally be defined as the angle between any two adjacent faces of a crystal. In crystallography, however, the interfacial angle to a crystal is the angle subtended between the normals drawn on the two faces concerned.

It has been observed that the interfacial angles of crystals of a particular mineral remain always constant. Since the atomic structure of the crystals of a particular mineral is fixed, the position of faces of such crystals will also be equal. So the corresponding interfacial angles are constant for all the crystals of a given mineral, provided they have identical chemical composition and are measured at the same temperature.

Law of constancy of interfacial-angle states that 'measured at the same temperature, similar angles on crystals of the same substance remain constant, regardless the size and the shape of the crystals.

[Contact-goniometers and reflecting-goniometer are used in measuring the interfacial angle of crystals.]

2. Law of rational-indices. Two crystals of the same substance may differ considerably in appearance that in number, size and shape of the individual faces. In order to describe the external form of crystals, a mathematical method of relating planes to certain imaginary lines in space is used.

The position of any plane can be uniquely fixed by the intercepts it makes on the axes of reference. The ratio of the distances
from the origin at which the crystal face cuts the crystallographic
axes, is known as the 'parameter' of a crystal face.

In the given figure, let OX, OY, OZ represents the crystallographic axes and ABC is a crystal face making intercepts of 'OA' on
'OX', 'OB' on OY and 'OC' on 'OZ'. Fig. given below.

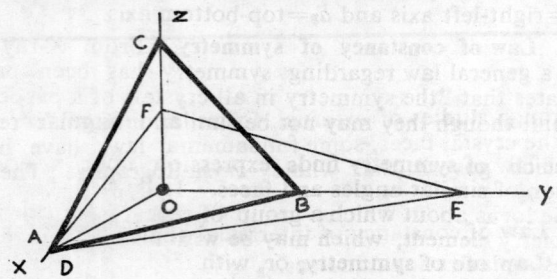

The parameter of the face ABC are given by the ratio of OA,
OB, OC. It is convenient to take the relative intercepts of this face
as standard length for the purpose of representing the position of any
other face such as DEF. In this case OD=OA, OE=2OB, OF=
½OC. Therefore the parameters are $\frac{1}{1}, \frac{2}{1}, \frac{1}{2}$ of DEF with reference
to the standard face ABC.

The reciprocals of the parameters are known as indices.

According to the crystallographic notation by Miller a law
has been established, which states that "the intercepts that any face
makes on the crystallographic axes are either infinite or small rational
multiples of the intercepts made by the unit form".

Hence the ratio between the intercepts on the axes of different
faces on a crystal can always be expressed by rational numbers as
1 : 2, 1 : 3, 1 : 4 ; but not as 1 : $\sqrt{2}$, 1 : $\sqrt{3}$ etc.

3. **Law of axial-ratio.** This law states that 'the ratio between
the lengths of the axes of the crystals of a given substance is
constant. This ratio is termed as 'axial-ratio'. Axial-ratio which is
the ratio of the lengths of the crystallographic expressed in terms of
one of the horizontal axes, usually, 'b'-axis, as unity.

In cubic system, where the three axes are identical, the ratio is
1 : 1 : 1 or $a : a : a$.

In tetragonal system, in case of zircon, $a : c = 1 : 0.6403$, in
case of rutile, $a : c = 1 : 0.64415$.

In Hexagonal system, in case of Beryl, $a : c = 1 : 0.4989$ and so
on.

4. Law of crystallographic axes. The position of the crystallographic axes are more or less fixed by symmetry of the crystals, for in most crystals they are symmetry axes or normal to symmetry plane. It has been observed that "crystals of a given mineral can be referred to the same set of crystallogrphic axes".

For example, all the crystals of galena may be referred to three crystallographic axes, which are of equal length, mutually perpendicular and are inter-convertible and are designated as a_1=front-back axis, a_2=right-left axis and a_3=top-bottom axis.

5. Law of constancy of symmetry. From X-ray studies of crystals, a general law regarding symmetry has been propounded ; which states that "the symmetry in all crystals of a particular species is constant, though they may not be similar in form".

The law of symmetry finds expression upon a crystal in the distrbution of similar angles and faces. It is well known that the geometric locus about which a group of repeating operations acts is the symmetry element, which may be with respect to a plane and is known as a plane of symmetry, or, with respect to a line, where it is said to be an axis of symmetry, or, with respect to a point, in which case it is known to be a centre of symmetry.

For example, the crystals of the mineral-'galena', whether it is octahedral, dodecahedral or cubic in shape, it shows the same symmetry elements, like, '9'-planes of symmetry, '13'-axes of symmetry and the centre of symmetry is also present. Similarly all the crystals of the mineral-Barite, shows '3'-planes of symmetry, 3-axes of two fold symmetry and the presence of centre of symmetry.

These aforesaid laws totally gevern all the aspects of crystallography.

CHAPTER 29

CRYSTAL SYSTEMS AND CLASSES

As we know, 'Crystallographic-Axes' are the imaginary lines passing through the centre of the crystal, but not lying in the same plane, and used as axes of reference for denoting the position of faces. They are usually axes of symmetry, normals to planes of symmetry or lines parallel to prominent edges of crystals. These crystallographic axes are of paramount significance in the classification of all the crystals into six major subdivisions, known as 'crystal-system'. The divisions of the crystals into systems are made on the basis of :

(*i*) number of crystallographic axes,

(*ii*) relative length of the crystallographic, and

(*iii*) angular relationship existing between the crystallographic axes.

Accordingly, crystals which may be referred to the same set of crystallographic axes, belong to the same crystal system. The systems, which have been recognised on this basis are as follows :

1. Isometric or Cubic System. $a_1=a_2=a_3$; $a \triangle a=90°$.

Here, there are three axes which are of equal length and mutually perpendicular, therefore interconvertible. Two axes and horizontal and the third one is vertical.

2. Tetragonal System. $a_1=a_2\neq c$, $a \wedge c=90°$.

In this system, there are three axes, of which two are of equal length and are horizontal, but both are at right angles to each other. The third one is vertical and it may be shorter or longer than that of the horizontal axes.

3. Hexagonal System. $a_1=a_2=a_3\neq c$; $a \wedge c=90°$.

Here, there are four axes of which three are in the horizontal plane, which are mutually inclind at 60°, but the angle between their positive ends is 120°. These horizontal axes are equal in length,

The fourth axis is vertical and may be shorter or longer in length than that of the horizontal ones.

4. Orthorhombic System. $a \neq b \neq c$; $a \wedge b \wedge c = 90°$.

It consists of three axes of unequal length and are mutually perpendicular.

5. Monoclinic System. $a \neq b \neq c$; $a \wedge c \neq 90°$; $b \wedge c = 90°$.

In this system, there are three unequal axes, which are designated as a, b and c. The axes 'a' and 'b' are lateral axes and the axes 'a' and 'b' and 'b' and 'c' makes 90° with each other but the axes 'a' and 'c' make an oblique angle with each other.

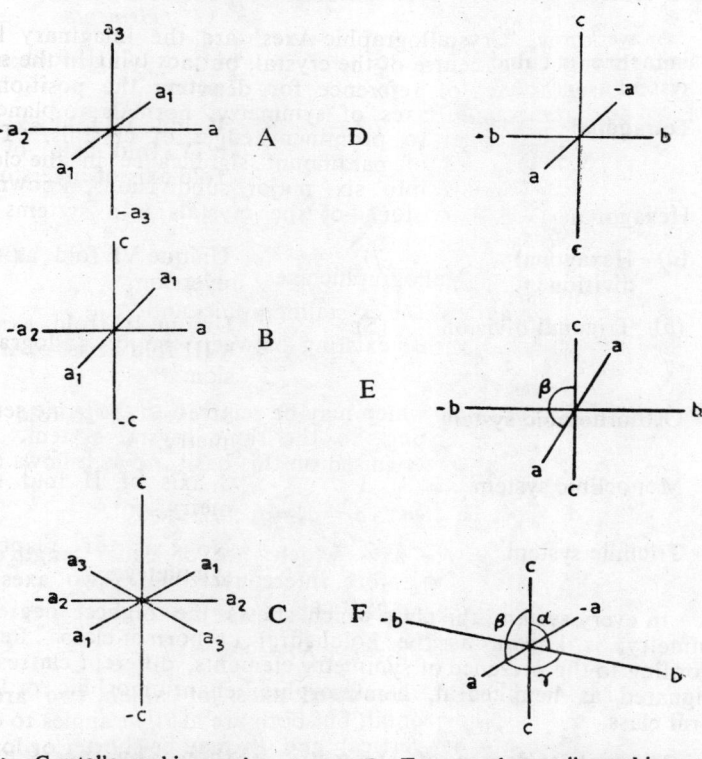

A—Crystallographic axes in isometric system,

B—Tetragonal crystallographic axes,

C—Hexagonal crystallographic axes,

D—Orthorhombic-axes,

F—Monoclinic axes,

F—Triclinic axes.

Fig. 29·1

6. Triclinic System. $a \neq b \neq c$; $a \wedge b \wedge c \neq 90°$.

Here, there are three unequal axes, which are inclined to each other at an oblique angle.

Classes. The classification of crystals into classes depends upon the degree and nature of the elements of symmetry. It has been shown mathematically that only thirty one combinations of symmetry elements are possible. A crystal may be without any symmetry. Thus there are in all only 32 crystal classes. All the 32 classes of crystals are grouped under six crystallographic systems, as follows :

System	No. of classes	Characteristic symmetry elements
1. Isometric or cubic system	5	4 axes of III fold symmetry.
2. Tetragonal	7	Unique IV fold axis or a IV fold axis of VII fold axis of inversion.
3. Hexagonal	12(7+5)	
(a) Hexagonal division	(7)	Unique VI fold axis or inversion.
(b) Trigonal division	(5)	Unique III fold axis or a III fold axis of inversion.
4. Orthorhombic system	3	3 axes of II fold symmetry.
5. Monoclinic system	3	1 axis of II fold symmetry.
6. Triclinic system	2	No axis or plane of symmetry.

In every system, the class which shows the highest degree of symmetry, is known as the holohedral or normal class. Besides according to the presence of symmetry elements, different classes are designated as hemihedral, hemimorphic, enantiomorphic or tetrahedral class.

The highest degree of symmetry is shown by the isometric system and the lowest degree of symmetry is exhibited by the triclinic system.

It has been observed from a statistical study of minera kingdom that more than 50% of the crystallised minerals belong tol the Monoclinic and Orthorhombic system. But, in general about 35%

of the minerals crystallize in the Monoclinic system ; about 27% in the Orthorhombic system ; about 15% in the cubic system ; about 10% in the Hexagonal system of which more than one-third belong to the Trigonal-division ; and 9% in the Triclinic and about 6% in the Tetragonal systems. In every system, majority of the minerals crystallises in the respective normal classes.

The various crystallographic systems and the different classes as included thereunder along with their symmetry elements and characteristic forms may be described as follows :

Isometric System

Crystallographic axes of reference. There are three axes of equal length which are mutually perpendicular and are interconvertible and are designated as 'a_1', 'a_2' and 'a_3'-axis.

a_1-axis runs front to back.

a_2-axis runs right to left.

a_3-axis runs top to bottom.

Unique Features of Isometric System :

(*i*) The axial ratios are 1 : 1 : 1.

(*ii*) Each of its classes has 4 axis of III fold symmetry.

(*iii*) Every form occurring in the system is a closed form and therefore can exist as a real crystal.

(*iv*) The forms occurring in this system are unique to it as the forms like pedion and pinacoid which occurs in all other systems do not occur in this system.

(*v*) It is the only system composed of classes containing more than one axis of 'n'-fold symmetry, where $n > 2$.

Classes Included in this System :

1. Narmal class, Holohedral class, Holosymmetric class or Hexaoctahedral class. It is also known as 'Galena-type' after the name of the mineral (Galena) which crystallises in this system.

2. Gyroidal class or plagiohedral class or entiomorphic class.

3. Hexa-tetrahedral class, Tetrahedral class.

4. Deploidal or pyritohedral class.

5. Tetartohedral class.

1. **Normal class.** Symmetry elements :

(*a*) 9-planes of symmetry { (*i*) 3-axial planes.

(*ii*) 6-diagonal planes.

(*b*) 13 axes of symmetry :

(*i*) 3-axis IV fold.

(*ii*) 4-axis III fold.

(*iii*) 6-axis II fold.

(*c*) Centre of symmetry present.

Forms present	No. of faces	Symbol
(*i*) Cube	6	(100)
(*ii*) Octahedron	8	(111)
(*iii*) Dodecahedron	12	(110)
(*iv*) Tetrahexahedron	24	210(*hko*)
(*v*) Trapezohedron or Tetragonal trisoctahedron	24	*hll* (*h>l*)
(*vi*) Trisoctahedron or Trigonal trisoctahedron	24	*hh*L(*h>l*)
(*vii*) Hexaoctahedron	48	*hkl*

In case of cube, each face is a 'square'; in octahedron each face is a 'equilateral triangle'; in dodecahedron each face is a 'rhombus'; in tetrahexahedron each face is a 'tetragonal pyramid'; in case of trapezohedron each face is a 'trapezium'; in trisocta-hedron each face is an 'isosceles tringle' and in hexaoctahedron each face is a 'sclene tringle.'

2. Gyroidal class. Symmetry Elements :

No plane of symmetry.

13 axes of symmetry :

(*i*) 3-IV fold axes.

(*ii*) 4-III fold axes.

(*iii*) 6-II fold axes.

No centre of symmetry.

Forms present. All the forms of the normal class are present in this class except the hexaoctahedron; in its place there occurs a face known as Gyroid, which has 24 faces, and its symbol is (*hkl*).

3. Hexatetrahedral class. Symmetry Elements :

6-planes of symmetry—all are diagonal.

7-axes of symmetry :

(*i*) 3-II fold axis.

(*ii*) 4-III fold axis.

No centre of symmetry.

Forms present	No. of faces	Symbol
(*i*) Tetrahedron	4	(111)
(*ii*) Trigonal-tristetrahedron	12	(211), *i.e.*, (*h l l*)
(*iii*) Deltohedron or Tetragonal-tristetrahedron	12	(*hhl*)
(*iv*) Hexatetrahedron	24	(*hkl*)

All the other forms of normal class are present here and the typical forms of this class are hemihedral forms.

4. Deploidal class. Symmetry Elements :

3-planes of symmetry.

7-axes of symmetry :

(*i*) 3-II fold axis.

(*ii*) 4-III fold axis.

Centre of symmetry present.

Forms present	No. of faces	Symbol
(*i*) Pyritohedron or Pentagonal dodecahedron	12	(*hko*)
(*ii*) Deploid or dyakisododecahedron	24	(*hkl*)

All the other forms of normal class are present.

5. Tetartohedral or Tetartoidal class. Symmetry Elements .

No plane of symmetry.

7-axes of symmetry :

(*i*) 3-II fold axis.

(*ii*) 4-III fold axis.

No centre of symmetry.

Forms present	No. of faces	Symbol
(*i*) Tetrahedron	4	(111)
(*ii*) Pyritohedron	12	(*hko*)
(*iii*) Trigonal tristetrahedron	12	(*hll*)

(*iv*) Deltohedron	12	(*hhl*)
(*v*) Tetartoid or Tetartohedral-penta-gonal-dodecahedron	12	(*hkl*)

Besides forms like cube and dodecahedron are present.

Tetragonal system. There are seven symmetry classes under this system as :

(*a*) Normal class, Ditetragonal dipyramidal class or Zircon type.

(*b*) Tetragonal trapezohedral class or trapezohedral class.

(*c*) Ditetragonal pyramidal or Hemimorphic class.

(*d*) Tetragonal-dipyramidal or Tripyramidal class.

(*e*) Tetragonal scalenohedral class or Sphenoidal class.

(*f*) Tetragonal pyramidal or Pyramidal-Hemimorphic class.

(*g*) Tetragonal disphenoidal or Tetartohedral class.

(*a*) **Normal Class.** Symmetry Elements :

    (1) 5-planes of symmetry :

        (*i*) 4-vertical (of which 2-diagonal and 2-axial).

        (*ii*) 1-horizontal plane.

    (2) 5-axes of symmetry :

        (*i*) 1-IV fold axis.

        (*ii*) 4-II fold (of which 2-axial and 2-diagonal).

    (3) Centre of symmetry present.

Forms present	*No. of faces*	*Symbol*
(1) Basal pinacoid	2	001
(2) Prism-Ist order	4	110
(3) Prism-2nd order	4	100
(4) Ditetragonal-prism	8	*hko*
(5) Tetragonal bipyramid (Ist order)	8	(*hhl*)
(6) Tetragonal bipyramid (2nd) order	8	(*hol*)
(7) Ditetragonal bipyramid	16	(*hkl*)

(*b*) **Tetragonal trapezohedral class.** Symmetry Elements :

    1. No plane of symmetry :

    2. 5-axes of symmetry.

        (*i*) 1-IV fold axis.

        (*ii*) 4-II fold axès.

    3.   No centre of symmetry.

Forms Present. All the forms of the normal class **except the** Ditetragonal bipyramid are present in this class, in its place there is a tetragonal trapezohedron, which contains eight faces and its symbol is (*hkl*).

    (*c*) **Ditetragonal pyramidal class.** Symmetry Elements :

        (1) 4-planes of symmetry.

        (2) 1-axis of IV-fold symmetry (vertical axis).

        (3) No centre of symmetry.

All the forms except the prisms of the normal class are represented by half the number of faces (of the corresponding forms). Therefore the typical forms of this class will be :

	Forms present	No. of faces	Symbol
(1)	Pedion	0·1	·001
(2)	Tetragonal pyramid (1st order)	4	*hhl*
(3)	Tetragonal pyramid (2nd order)	4	*hol*
(4)	Ditetragonal pyramid	8	*hkl*

    (*d*) **Tetragonal-dipyramidal class.** Symmetry Elements :

        (*i*) 1-plane of symmetry (horizontal).

        (*ii*) 1-axis of IV-fold symmetry (vertical).

        (*iii*) Centre of symmetry present.

	Forms present	No. of faces	Symbol
(*i*)	Tetragonal-prism IIIrd order	4	*hko*
(*ii*)	Tetragonal-bipyramid IIIrd order	8	(*hkl*)

Along with the above two typical forms, all the other forms of normal classes present (except the ditetragonal prism and the ditetragonal bipyramid). As this class consists of pyramids of 1st, 2nd and 3rd order, it is known as Tripyramidal class.

    (*e*) **Tetragonal scalenohedral class.** Symmetry Elements :

        (*i*) 2-vertical-diagonal planes of symmetry.

(*ii*) 3-axes of II fold symmetry.

(*iii*) No centre of symmetry.

All the forms of the normal class are present here, except the tetragonal bipyramid and ditetragonal bipyramid, which are substituted by the following two typical forms :

Forms present	No. of faces	Symbol
(*i*) Tetragonal sphenoid	4	(*hhl*)
(*ii*) Tetragonal scalenohedron	8	(*hkl*)

(*f*) **Tetragonal pyramidal class.** Symmetry Elements :

(*i*) No plane of symmetry.

(*ii*) 1-vertical axis of IV-fold symmetry.
(*iii*) No centre of symmetry.

Forms present	No. of faces	Symbol
(*i*) Pedion	1	001
(*ii*) Tetragonal prisms :		
1st order	4	110
2nd order	4	100
3rd order	4	*hko*
(*iii*) Tetragonal pyramid :		
1st order	4	*hhl*
2nd order	4	*hol*
3rd order	4	*hkl*

In this class, all the forms excepting the prisms are represented by half the number of faces of the corresponding forms of normal class, and in the 3rd order pyramid it is in respect of the form of the tetragonal-dipyramidal class.

(*g*) **Tetragonal disphenoidal class.** Symmetry Elements :

(*i*) No plane of symmetry.

(*ii*) 1-axis of rotary inversion of four-fold symmetry.

(*iii*) No centre of symmetry.

Forms present	No. of faces	Symbol
Tetragonaldisphenoid		
1st order	4	*hhl*
2nd order	4	*hol*
3rd order	4	*hkl*

Other forms like tetragonal prism of 1st, 2nd and 3rd order and basal pinacoid are present.

Hexagonal system. This system consists of the highest number of symmetry classes, *i.e.*, 12 classes. They are conveniently grouped into two divisions, namely :

(1) Hexagonal-division.
(2) Trigonal-division.

Seven symmetry classes have been included in the Hexagonal division and five are grouped in the Trigonal division.

Hexagonal division. The seven symmetry classes as included under this division are as follows :

1. Normal class or Dihexagonal dipyramidal class.
2. Hexagonal trapezohedral class.
3. Dihexagonal pyramidal class.
4. Ditrigonal dipyramidal class.
5. Hexagonal dipyramidal class.
6. Hexagonal pyramidal class.
7. Trigonal dipyramidal class.

(1) Normal class. Symmetry Elements :

(*i*) 7-planes of symmetry :

1-horizontal, 6-vertical planes of which three are axial and three diagonal.

(*ii*) 7-axes of symmetry :

1-VI fold axis of symmetry, 6-II fold axes of symmetry.

(*iii*) Centre of symmetry present.

Forms present	No. of faces	Symbol
(a) Basal pinacoid	2	0001
(b) Prism of the 1st order	6	$10\bar{1}0$
(c) Prism of the 2nd order	6	$11\bar{2}0$
(d) Dihexagonal prism	12	$hk\bar{i}o$
(e) Bipyramid-1st order	12	$(ho\,\bar{h}\,l)$
(f) Bipyramid-2nd order	12	$(hh\,\overline{2h}\,l)$
(g) Dihexagonal bipyramid	24	$(hk\,\bar{i}\,l)$

In the above case, '$h > k$' and '$h+k = -i$.'

2. Hexagonal trapezohedral class. Symmetry Elements :

    (*i*) No plane of symmetry.

    (*ii*) 7-axes of symmetry :

        1-vertical axis of VI fold symmetry', 6-axes of II fold symmetry.

    (*iii*) No centre of symmetry.

All the forms of normal class are present here, excepting the dihexagonal bipyramid, in its place, there occurs a 'hexagonal trapezohedron' having '12' faces and its symbol is $(hk\ i\ l)$.

3. Dihexagonal pyramidal class. Symmetry Elements :

    (*i*) 6-planes of symmetry.

    (*ii*) 1-axis of VI-fold symmetry.

    (*iii*) No centre of symmetry.

All the forms of the normal class are present in this class, but they are represented by half the number of faces of the corresponding forms of the normal class.

4. Ditrigonal dipyramidal class. Symmetry Elements :

    (*i*) 4-planes of symmetry, of which one is horizontal and three are vertical.

    (*ii*) 4-axes of symmetry :

        1-axis of III fold symmetry, 3-axes of II fold symmetry.

    (*iii*) No centre of symmetry.

Forms present	No. fo faces	Symbol
(*a*) Trigonal prism	3	$10\bar{1}0$
(*b*) Trigonal pyramid	6	$10\bar{1}1$ $(ho\,\bar{h}l)$
(*c*) Ditrigonal prism	6	$hk\bar{i}o$
(*d*) Ditrigonal pyramid	12	$(hk\,\bar{i}l)$

Excluding these forms, other forms of the normal class like hexagonal prism and pyramid of second order and basal pinacoid also occur in this class.

5. Hexagonal dipyramidal class. Symmetry Elements :

    (*i*) *l*-plane of symmetry (horizontal plane).

    (*ii*) *l*-axis of VI fold symmetry.

    (*iii*) Centre of symmetry present.

In this class, all the forms of the normal class are present excepting the dihexagonal prism and dihexagonal bipyramid ; in.

their place two other forms are present as hexagonal prism-IIIrd order (6 faces with symbol $hk\bar{i}o$) hexagonal pyramid of IIIrd order (12 faces with symbol $hk\,i\,l$), respectively.

6. Hexagonal pyramidal class. Symmetry Elements :

(*i*) No plane of symmetry.

(*ii*) 1-axis of VI-fold symmetry (vertical).

(*iii*) No centre of symmetry.

Here, in this class, forms like hexagonal prism 1st, 2nd and 3rd order and hexagonal pyramids of 1st, 2nd and 3rd order and the pedion is present. But, it is to be kept in mind that except the pedion all the other forms are represented by six-faces.

7 Trigonal dipyramidal class. Symmetry Elements :

(*i*) 1-plane of symmetry (horizontal).

(*ii*) 1-axis of III fold symmetry (vertical).

(*iii*) No centre of symmetry.

The characteristic forms of this class are the three types of trigonal. prisms and the three types of trigonal pyramids :

(*a*) *Trigonal prisms* :

1st order	3 faces	$10\bar{1}0$
2nd order	3 faces	$11\bar{2}0$
3rd order	3 faces	$hk\bar{i}o.$

(*b*) *Trigonal pyramids* :

1st order	6 faces	$(ho\,\bar{h}\,l)$
2nd order	6 faccs	$(hh\,\overline{2h}\,l)$
3rd order	6 faces	$(hk\,i\,\bar{l})$

(*c*) *Basal pinacoid* :

	2 faces	0001.

Note. The characteristic symmetry of hexagonal division is 1-axis of VI-fold symmetry or rotary inversion axis of VI fold symmetry.

Trigonal division. All the crystals of the trigonal division are characterised by a vertical axis of three-fold symmetry. This division includes five symmetry classes as follows :

1. Rhombohedral or Hexagonal Scalenohedral class.

2. Trigonal-trapezohedral class.

3. Ditrigonal pyramidal class.

  4.   Tri-rhombohedral class.

  5.   Trigonal-pyramidal class or tetartohedral class.

  1. Rhombohedral or Normal class.   Symmetry Elements :

   (*i*) 3-vertical, diagonal planes of symmetry.

   (*ii*) 4-axes of symmetry, of which 1-axis is of three-fold
         symmetry, and the three horizontal axes are of II fold
         symmetry.

   (*iii*) Centre of symmetry present.

  In this class, all the forms of the hexagonal division are present
except the hexagonal pyramid of 1st order and dihexagonal bipyramid ; instead two other forms present are as follows :

  (*a*) Rhombohedron   6 faces   (*ho h̄ l*)
  (*b*) Scalenohedral   12 faces   (*hk ī l*)

  2. Trigonal trapezohedral class.   Symmetry Elements :

   (*i*) No plane of symmetry.

   (*ii*) 4-axis of symmetry, of which one axis is of III fold
         symmetry and the rest three are of II fold symmetry.

   (*iii*) No centre of symmetry.

  The characteristic form of this class is the trigonal-trapezohe-
dron, which has six faces and the symbol is (*hk i l*). Besides,   there
are hexagonal prism 1st order ; trigonal prism (*hh 2h̄ o*) having three
faces ; Ditrigonal prism-6 faces (*hk ī o*), Rhombohedron, 1st order
(6 faces, *ho h̄ l*), Trigonal bipyramid 2nd order-6 faces (*hh 2h̄ l*) ; and
basal pinacoid (0001).

  3. Ditrigonal pyramidal class.   Symmetry Elements :

   (*i*) 3-vertical, diagonal planes of symmetry.

   (*ii*) 1-vertical axis of III fold symmetry.

   (*iii*) No centre of symmetry.

Forms present	*No. of faces*	*Symbol*
(*a*) Pedion	1	0001
(*b*) Trigonal prism 1st order	3	10 ī 0
(*c*) Hexagonal prism (2nd order)	6	*hh 2h̄ o*
(*d*) Ditrigonal prism	6	*hk ī o*
(*e*) Hexagonal pyramid	6	
(hemimorphic) (2nd order)		(*hh 2h̄ l*)

    (*f*) Trigonal pyramid (hemimorphic) 3   ($ho\,\bar{h}\,l$)

    (*g*) Ditrigonal pyramid   6

        (hemimorphic)   ($hk\,\bar{i}\,l$)

These are the forms present in this class.

4. Trirhombohedral class. Elements of symmetry :

    (*i*) No plane of symmetry.

    (*ii*) 1-vertical axis of III fold symmetry.

    (*iii*) Centre of symmetry present.

This class consists of forms like basal pinacoid and hexagonal prisms of 1st, 2nd and 3rd order, which are having 6-face each and there are also rhombohedrons of 1st, 2nd and 3rd order having six-faces each. Pyramids are not occurring in this class.

5. Trigonal pyramidal class. Symmetry Elements :

    (*i*) No plane of symmetry. ˙

    (*ii*) 1-axis of III-fold symmetry.

    (*iii*) No centre of symmetry.

All the forms occurring in this class are hemimorphic. They are :

(*a*) *Trigonal pyramid*	No. of faces	Symbol
1st order	3 face	($ho\,\bar{h}\,l$)
2nd order	3 face	($hh\,\overline{2h}\,l$)
3rd order	3 face	($hk\,\bar{i}\,l$)
(*b*) *Trigonal prism* :		
1st order	3 face	$10\,\bar{1}\,0$
2nd order	3 face	$11\,\bar{2}\,0$
3rd order	3 face	$hk\,\bar{i}\,o$
(*c*) *Pedion* :		
	1 face	0001

Orthorhombic system. It is customary to orient a **crystal** of this system in such a way that '*b*'-axis is greater than '*a*'-axis. **There** is a divergence of opinion regarding the relative length of the '*c*'-axis ; some crystallographers follow the convention $c{>}a{>}b$, whereas other prefer $c{>}b{>}a$. But all choose the length of **the** '*b*'-axis as unity. In this system the '*b*'-axis is known as 'Macro axis' and the '*a*' axis as 'Brachy axis'. There are three **symmetry** classes in this system as follows :

1. Normal class or Orthorhombic dipyramidal class.
2. Orthorhombic pyramidal class.
3. Orthorhombic disphenoidal class.

1. Normal class. Symmetry Elements :

(a) Plane of symmetry=3 axial plane of symmetry.
(b) Axis of symmetry=3 axis of II fold symmetry.
(c) Centre of symmetry=Present.

Forms present	No. of faces	Symbol
(i) Basal pinacoid	2	001
(ii) Brachy pinacoid	2	010
(iii) Macro pinacoid	2	100
(iv) Prism	4	110, i.e., hko
(v) Brachy dome	4	okl
(vi) Macro dome	4	hol
(vii) Bipyramid	8	hkl

2. Orthorhombic pyramidal class. Symmetry Elements :

(a) 2-planes of symmetry.
(b) 1-axis of II fold symmetry.
(c) No centre of symmetry.

Forms like brachy-pinacoid, macro-pinacoid and prisms are occurring in this class and other forms like basal pinacoid, brachy-dome, macro-dome and pyramids are represented by half the number of faces of the corresponding forms of the normal class.

3. Orthorhombic disphenoidal class. Symmetry Elements :

(a) No plane of symmetry.
(b) 3-axes of II fold symmetry.
(c) No centre of symmetry.

All the forms of the normal class except the bipyramid are present in this class. In place of the bipyramid, there occurs a disphenoid, consisting of four faces with the symbol (hkl).

Monoclinic system. Of the three crystallographic axes of the Monoclinic system, the 'b' axis known as the 'Ortho-axis,' and the 'a'-axis which runs up and away from the observer is known as the 'clino-axis'.

This system consists of three symmetry classes as :

(1) Normal or Prismatic class.　(2) Dometic class.
(3) Sphenoidal class.

1. Normal class. Symmetry Elements :

   (*i*) 1-plane of symmetry.

   (*ii*) 1-axis of II-fold symmetry.

   (*iii*) Centre of symmetry present.

Forms present	No. of faces	Symbol
(*a*) Basal pinacoid	2	001
(*b*) Ortho pinacoid	2	100
(*c*) Clino pinacoid	2	010
(*d*) Prism	4	*hko*
(*e*) Ortho dome	2	*hol*
	(*llrl* to *b*-axis)	
(*f*) Clino dome	4	*okl*
(*g*) Pyramid (hemimorphic)	4	*hkl*.

2. Domatic or Clinohedral class. Symmetry Elements :

   (*i*) 1-plane of symmetry (ac-plane).

   (*ii*) No axis of symmetry.

   (*iii*) No centre of symetry.

Forms present	No. of faces	Symbol
(*a*) Basal pinacoid	1	001
(*b*) Ortho pinacoid	1	100
(*c*) Clino pinacoid	2	010
(*d*) Ortho dome	1	*hol*
(*e*) Hemi clinodome	2	*okl*
(*f*) Prism	2	*hko*
(*g*) Pyramid	2	*hkl*.

3. Sphenoidal or Hemimorphic class. Symmetry Elements :

   (*a*) No plane of symmetry.

   (*b*) 1-axis of II fold symmetry.

   (*c*) No centre of symmetry.

Excepting the forms like Basal pinacoid, Orthopinacoid and ortho dome, other forms are represented by half the number of faces of the corresponding form of the normal class. They are as follows :

Forms present	No. of faces	Symbol
(*i*) Sphenoid	2	*hkl*
(*ii*) Prism	2	*hko*

| (*iii*) Clino pinacoid | 1 | 010 |
| (*iv*) Clino dome | 2 | *okl*. |

The pyramid of the normal class, is substituted in this class by the sphenoid.

Triclinic system. For a crystal of this system, the orientation can be a matter of choice. To orient the crystal, any axis is placed vertically, the longer lateral axis is made to extend downwards to the right and the front-rear axis downward to the front. The angle between '*a*' and '*b*' axis is denoted by 'Ψ', between '*a*' and '*c*' axis by 'β' and between '*b*' and '*c*' by '*a*'. Usually '*a*' and '*b*' axis are so choosen that the '*a*'-axis becomes the shorter axis and is known as 'Brachy-axis', similarly the longer '*b*'-axis is known as 'Macro-axis'.

This system includes two symmetry classes as :

1.   Normal or Pinacoid class.
2.   Asymmetric or Hemihedral or Pedial class.

1. Normal class. Symmetry Elements :

   (*a*) No plane of symmetry.
   (*b*) No axis of symmetry.
   (*c*) Centre of symmetry present.

All the forms of this class are pinacoids. Each form of the class includes two faces parallel to one another and symmetrical with reference to the centre of symmetry. The forms present in this class are the basal pinacoid, brachy-pinacoid, ortho-pinacoid, prism, macrodome, brachydome and pyramids.

2. Pedial class. Elements of symmetry :

   (*a*) No plane of symmetry.
   (*b*) No axis of symmetry.
   (*c*) No centre of symmetry.

In this class all the forms are pedions, *i.e.*, the forms are represented by one face only. Hence this class is known as Pedial-class.

CHAPTER 29

CRYSTAL HABITS AND TWINNING

The crystal habit of a mineral is a term used to denote the relative development of faces and forms in its crystals; in other words, it refers to the form or combination of forms commonly occurring in crystals of that mineral.

When the external form of a crystal is column-like, bounded by a group of vertical faces, it is said to have a *prismatic habit*. Similarly, when a crystal displays maximum development of pyramidal faces, It is said to possess *pyramidal habit* and pyramidal crystals usually taper at two opposite ends. Whenever a crystal shows prominent flat surfaces because of the development basal plane the habit of the crystal becomes *tabular*.

Crystals occur in all possible sizes from the merest microscopic to crystals measuring in tens of metres in diameter. The variations in habit is due to the significant variations in the conditions attending crystallisation, including pressure, temperature, quality and/or quantity of foreign material.

It is usually observed that crystals of a particular mineral shows similar crystal habits, but in certain case, even though there are same faces, there is a difference in crystal habit, which is mainly because of the differential development of the faces.

Twinning. Two or more crystals of the same or related mineral species may sometimes intergrow in such a way that the individual parts are in reverse position to each other. Such crystals are called '*twinned crystals*'. Sometimes, it also appears as if one-half of the twinned crystal is produced by rotation of 180°, about some crystal axis common to both.

Twinned crystals are characterised by '*re-entrant*' angles, and there are cleavages in different directions in different parts of what is apparently the same crystal.

(*a*) **Twin-plane.** It is an imaginary plane which divides a twin-crystal into two halves such that one-half is a reflection of the other.

(*b*) **Twin-axis.** It is an imaginary line about which rotation is necessary to bring the twin to its untwinned state.

Twinning-centre. When twinning is defined with respect to symmetry about a point, the referred point is called the 'twin-centre.'

(*d*) **Composition plane.** The plane by which the component crystals of a twin are joined is called the composition plane.

A twin plane is always a possible crystal face and the twin axis is normal to it. The composition plane usually but not necessarily, coincides, with the twin plane ; when not coincident, the two planes are mutually perpendicular.

Types of twins :

(*i*) **Contact twin.** It consists of two halves united by composition plane such that one part is the mirror image of the other, *e.g.*, 'Orthoclase'.

(*ii*) **Penetration twin.** Here the twin appears as if the individuals were crossing each other. Staurolite, pyrite, fluorite etc. show this type of twinning.

In case of pyrite, this twinning called 'Iron-Cross'.

(*iii*) **Repeated twin :**

(*a*) *Polysynthetic or Lamellar twins.* Here the twin planes are parallel to each other, *e.g.*, plagioclase feldspars, calcite etc.

(*b*) *Symmetrical or Cyclic twin.* Here the twin plane does not remain parallel but tends to produce circular forms as in spinel, rutile etc. Such twinning in parts as apparent symmetry of higher grade than that of the individuals and accordingly this is also known as *mimetic-twin.*

e.g., Aragonite appears to be hexagonal.

In case of rutile, the cyclic twin is known as *'Geniculate-twin'.*

In case of Gypsum, it is known as *'Butterfly-twin.*

Important Minerals and their Characteristic Twins :

1. Carlsbad twin	Twin plane (010) *i.e.*, clinopinacoid	Orthoclase and plagioclases
2. Baveno twin	Twin plane (021) *i.e.*, clino dome	—do—
3. Manebach twin	Twin plane (001) *i.e.*, basal pinacoid	—do—
4. Swallo-tails	Twin plane (100) *i.e.*, orthopinacoid	Gypsum

5. Pericline twin	Composition plane is (010), and *b*-axis is the twin axis	Plagioclase
6. Albite-twin or Polysynthetic twin	Twin plane is (010), *i.e.*, the brachy pinacoid	Plagioclase
7. Cyclic-twin or Mimetic twin	Twin plane is (110), *i.e.*, prism face.	Aragonite
8. Maltese-cross	Twin plane is (*okl*), *i.e.*, brachydome	Staurolite
9. Skew-twin	Twin plane is the pyramidal face, *i.e.*, (*hkl*)	Staurolite
10. Geniculate twin	Twin face is (101)	Rutile
11. Iron-cross	Octahedron face is the twin plane	Pyrite.

Thus twinning serves as a useful means of identifying certain minerals.

CHAPTER 30

MINERALOGY

1. Optical mineralogy.
2. Properties of minerals.
3. Study of mineral groups.

—Feldspar., —Quarz,
—Amphiboles. —Pyroxenes,
—Chlorites. —Micas,
—Garnets, —Carbonates.

OPTICAL-MINERALOGY

Principles of optics. In the context of optical mineralogy, it includes refraction, double-refraction and total internal reflection.

(*i*) **Refraction.** As we know, when light passes from one medium to another, there is, in general, an increase or decrease in its velocity results and accordingly there is also a change in the direction of its propagation. When light travels from a rarer to a denser medium, it is bent towards the normal on entering the latter and *vice versa*.

Refractive index is a ratio between the sine of the angle of incidence and the sine of the angle of refraction, which is always a constant for the two media concerned. The index of refraction depends upon the nature of the substance and the kind of light used and increases with density and decrease in wavelength. Thus, since the wavelength of red light is greatest and that of the violet light is the least, the index of refraction for the violet light is greater than that of the red. The difference of R.I. between red and violet light is called 'Dispersion'.

Methods of determining the refractive index. There are several methods of determining the refractive indices of minerals as follows :

(*a*) With test liquids (Immersion method),
(*b*) By Prism methods,

(c) By Refractometers, and

(d) By Polarising Microscope.

In case of determining the refractive index by polarizing microscope, the relative index of refraction of a mineral with respect to the immersing medium—The Canada Balsam ; or the two adjoining minerals in contact is determined. The methods commonly employed, are as

(a) Central illumination or Becke-line method.

(b) Oblique illumination method.

(a) **Becke-line method.** The principle involved is the total reflection of light incident at more than the critical angle when passing from a mineral of greater index to that of a lesser index in thin section. Accordingly a portion of the beam is deflected towards the mineral with greater index which results in a thin band of light visible just inside the boundary of greater index. This band of light is called the Becke-line, which moves towards the mineral of greater refractive index when the microscope-tube is slightly raised and the reverse effect is produced when the tube is lowered.

(b) **Oblique-illumination method.** Oblique illumination is made by cutting off half of light by using a finger or card below the stage. Thus half of the field remains illuminated and one side of the mineral will be dark and the opposite side lighted. It has been observed that of the rays pass from a mineral of higher index into a mount of lower index, they are concentrated by refraction and form a light band, if they pass from the mount of lower index into the higher mineral they are spread out by refraction and so produce a shadow. In general "if the shadow appears on the side away from the dark half, the refractive index of the mineral in question is greater than that of the adjoining medium and *vice versa*"

(ii) **Birefringence.** Excepting minerals which crystallize in the isometric system or are amorphous all the minerals are anisotropic *i.e.*, a ray of light striking the surface normally or obliquily, breaks into two rays vibrating along planes which are perpendicular to each other. These two rays travel with different velocities and are differently refracted. Both of these rays are plane polarized, in other words they vibrate in definite directions or in definite planes. Such a phenomenon whereby there occurs a division of the refracted light into two rays is called '*Double Refraction*'.

If a transparent calcite rhomb is placed over a dot, two images of the dot will be seen while observing from upwards. Moreover on rotating the crystal one of the images remain stationary, while the other moves around the stationary image. The stationary image is called the 'ordinary image', produced by the ray called ordinary ray which has passed through the crystal as if it were an isotropic medium. The other is known as the 'Extra-ordinary' image produced by the Extra-ordinary ray.

The extraordinary ray vibrates in a plane containing '*c*'-crystallographic axis, *i-e.*, within the principal section and the ordinary ray vibrates perpendicular to the principal section.

In every anisotropic mineral there is a direction in which both the extraordinary and ordinary ray travels with the same velocity. This direction is called the '*Optic axis*'. In an anisotropic mineral the velocity of ordinary ray is constant for all directions but that of the extraordinary ray varies with direction, becoming minimum or maximum at right angles to the optic axis.

Birefringence is a measure of the difference between the maximum and minimum refractive indices of a particular mineral ; in other words, it is the difference between the refractive indices of the two-rays, *i.e.*, extraordinary and ordinary ray. Calcite shows the highest birefringence, which is 0·142, because the refractive indices of the extraordinary and ordinary rays are 1·516 and 1·658 respectively.

According to the number of directions along which no double refraction occurs, anisotropic minerals are classified into two groups :

(*a*) *Uniaxial minerals.* Minerals of teragonal and hexagonal system, which possess only one optic axis.

(*b*) *Biaxial minerals.* Minerals of the orthorhombic, monoclinic and triclinic system possess two optic axes, along which no double refraction occurs.

Birefringence also determines the optic sign of the minerals. "In any mineral if the extraordinary ray is faster than the ordinary ray, *i.e.*, the refractive index of the ordinary ray is more than that of the extraordinary ray, the mineral is said to be negative." Similarly when the ordinary ray is fast, *i.e.*, the refractive index of the extraordinary ray is greater than that of the ordinary ray, the mineral is said to be positive. Thus Calcite is optically negative and Quartz is optically positive.

Birefringence of some common minerals are as follows :

Quartz   =0·009,   Orthoclase =0·008,

Kyanite   =0·016,   Tourmaline=0·02,

Zircon   =0·06.

Birefringence is used in determining the thickness of the section.

It has also been observed that greater the birefringence, higher the interference colour.

The minerals showing double refraction are also called as birefringent minerals.

Nicol prism. This is a device to produce and analyze polarized light through the principle of double refraction. It consists of a rhombohedron of calcite whose length is three times the breadth. The top and bottom surfaces are ground down to give an angle of 68° with the long edge. The block is then cut along the smaller diagonal and cemented together again with the help of a layer of Canada Balsam whose refractive index is 1·537, which acts as a rarer medium for an ordinary ray (R.I. 1·658) and a denser medium for the extraordinary ray (1·516). Since the ordinary ray travels from denser to rarer medium, it is totally internally reflected, as it is so arranged that the angle of incidence at the Canada Balsam layer is greater than the critical angle for the Ordinary ray. But in case of extraordinary ray, since it travels from rarer to denser medium it passes through the nicol with little deviation and vibrates parallel to the short-diagonal of the crystal.

These are called nicol prisms which are used in the petrological microscopes both for producing and analyzing the polarised light and thus aids in the identification of minerals.

PLEOCHROISM

It is an important optical properties of minerals in thin-section, in which the change in quality and quantity of colour is observed on rotation of the stage through an angle of 90°.

The colour shown by a mineral in this section results from the absorption of certain 'colours' (wavelengths) from the incident white light ; the resulting transmitted light being complementary in colour to that absorbed. Thus pleochroism is defined as the variation in colour resulting from differential absorption of wavelength in different directions.

1. Isotropic substances. In isotropic substances, the absorption of light is same in every direction and there is no variation of intensity and colour of the light during the rotation of the stage. Accordingly isotropic substances are non-pleochroic.

2. Anisotropic substances. These minerals show distinct pleochroism, but it depends on the crystallographic orientation of the section cut from the mineral. Different sections of the same mineral show different degree of pleochroism.

(*a*) **Uniaxial minerals.** Basal sections of (tetragonal and hexagonal) uniaxial minerals are isotropic ; only prismatic sections show pleochroism. The greatest degree of pleochroism occurs when the crystallographic direction of the section are either parallel or perpendicular to the vibration plane of the polarizer, in other words, coincides with those of the ordinary and extraordinary rays. Thus the pleochroism is seen only in two directions and accordingly minerals are said to be '*dichroic*.'

(*b*) **Biaxial minerals.** If a section of biaxial mineral, cut at right angles to an optic axis be examined ; it will be found that the absorption is the same in every direction, and it will behave as an isotropic section.

In other sections of the biaxial minerals, the variation of colour occurs according to the three optic directions 'X, Y and Z' ; of which 'X' is the fastest and 'Z' is the slowest direction of transmission of light. Thus absorption is three-fold in biaxial minerals and biaxial minerals are, therefore, known as trichroic.

It should be noted that pleochroism is the property of coloured minerals only but all coloured minerals are not pleochroic. In colourless minerals there is no question of pleochroism.

Pleochroic haloes. These are circular areas around minute inclusions, which are darker coloured than the remainder of the crystal and commonly exhibit pleochroism. The inclusions, in these cases, are the product of radioactive disintegration and themselves are radioactive and produce the ionisation effect on the surrounding portion.

Pleochroic haloes are observed in muscovite, biotite, tourmaline, cordierite, andalusite, pyroxene, amphibole etc. and the included minerals belong to zircon, apatite, rutile, sphene etc. These haloes, however, are destroyed above 500° and thus help in determining the upper limit of formation of such minerals.

EXTINCTION

When the vibration directions of the ordinary and extraordinary rays of an anisotropic mineral are parallel to the vibration directions of the nicols in a petrological microscope, no light reaches the eye and the mineral is said to be in extinction. This is because of the fact that light which can pass the polarizer also passes the mineral but is stopped by the analyser, as it has a vibration direction perpendicular to that of the polarizer. This phenomenon occurs four times during the rotation of the stage of the microscope (the thin section) through an angle of 360° under crossed-nicols position.

An isotropic mineral is always in the position of extinction between crossed nicols. In case of anisotropic mineral, the two adjacent positions of extinction are separated from each other by 90°.

There are four types of extinction as follows :

(*i*) **Straight extinction.** If the section becomes darkest when its length (which is usually parallel to *c*-axis) or a prominent cleavage direction is parallel to any of the cross-wires, the extinction is straight or '0°'.

(*ii*) **Oblique or Inclined extinction.** When a mineral is in extinction with the cleavage or crystal boundaries lying oblique to the planes of vibration of the two nicols, *i.e.*, inclined to the cross-wires, the extinction is said to be 'Inclined or Oblique.' All the Biaxial minerals (Orthorhombic, monoclinic, triclinic) show oblique extinction.

(*iii*) **Symmetrical extinction.** This type of extinction is shown by minerals with squarish outline or rhombic cross-section. In this case the mineral section extinguishes with the planes of vibration of the nicols being parallel to the diagonals of the rhombic pattern.

(*iv*) **Wavy or Undulose extinction.** In this case it appears as a band or series of bands of darkness crossing a single crystal unit during rotation. It may be owing to strain in mineral, that the mineral fail to extinguish completely. Quartz often shows wavy extinction.

Extinction angle. The angle between a crystallographic direction and the position of maximum extinction is called the extinction angle. Ordinarily the angle is measured with reference to cleavage direction but in the absence of cleavage, crystal outline, prominent crack, twin plane etc. are taken into account for the measurement of extinction angle.

Interference colour. Anisotropic minerals in positions intermediate between the positions of extinction show some colours which is the outcome of the interference of the two rays of light having the same vibration plane (after their emergence from the analyzer) but one is retarded with reference to the other ; accordingly these colours are known as interference colour.

These colours depend upon the thickness of the mineral section, birefringence of the section, crystallographic orientation of the section etc.

[**Note.** Ordinarily a section is 0·03 mm thick].

CHAPTER 31

SIMPLE POLARISING MICROSCOPE

It is an optical device used to study minerals and rocks in thin sections. It makes use of polarised light for the puspose of identification of minerals. It is equipped with the following mechanical and optical parts:

1. Mechanical parts :

(a) Base or foot.

(b) Pillar, attached to the base.

(c) Joint for making the upper part of the microscope inclined.

(d) Arm, attached to the upper end of the pillar.

(e) Tube, attached to the upper part of the arm, about 10″ in length with slots for inserting certain optical devices, such as mica plate, gypsum plate, quartz-wedge etc.

(f) Stage, which is a circular disc with a hole in the centre, above which the mineral section, *i.e.*, the slide is clipped.

(g) Adjusting screws, the tube can be moved up or down by means of two screws known as Coarse and Fine—adjusting screw.

(h) Mirror Arm and Fork to rotate and turn the mirror.

2. Optical parts :

(a) Mirror, one side plane and the other side concave to reflected light.

(b) Lower Nicol-prism, *i.e.*, polariser to produce plane polarised light.

(c) Iris Diaphragm lessens illumination of field, if slightly closed.

(d) Condenser, to increase the intensity of light.

(e) Objectives, two or more lenses of different focal lengths, used to magnify the image of objects.

(f) Upper-nicol (Analyser) or the Upper Nicol Prism used to analyze the plane of vibration of light which passes through the mineral section.

(g) *Eye-piece or Ocular.* It is with cross wires at right angles and a system of lenses fitted into the upper end of the tube, which magnifies the image of objects.

(h) *Bertrand lens.* It is used to magnify interference figures.

Optical Accessories :

(a) **Quartz wedge.** It is used for the determination of interference colour, optical sign and nature of vibration-direction of the mineral.

(b) **Mica plate.** Also known as quarter-wave plate or glimmer plate, which gives a pale neutral gray interference colour, when put into the slot of the tube, used for determining the optical sign of a mineral.

(c) **Gypsum plate.** Also known as sensitive-tint plate, as it produces sensitive violet red interference colour of first order ; used for determining the optical sign of a mineral.

(d) **Bereck compensator.** An optical device made of calcite.

Through the examination by a Petrological-microscope, the following characteristics of a mineral section are identified.

(i) Colour.

(ii) Form of a mineral, *i.e.*, whether euhedral, subhedral or anhedral.

(iii) Cleavage.

(iv) Relief of a mineral.

(v) Refractive index

(vi) Inclusions of any other substance.

(vii) Alteration.

($viii$) Pleochroism.

(ix) Interference colour.

(x) Extinction.

(xi) Extinction angle.

(xii) Twinning.

CHAPTER 32

PHYSICAL PROPERTIES OF MINERALS

Minerals generally occur in aggregates of more or less imperfectly developed crystals. 'Structure' is the usual term used to denote the state of aggregation or shape of the minerals, which may be of the following type :

(*i*) **Crystalline.** When the minerals are in the form of imperfectly developed crystals.

(*ii*) **Amorphous or massive.** When the mineral does not possess crystalline form.

(*iii*) **Earthy.** It is a uniform aggregate of exceedingly minute particles like those of china-clay.

(*iv*) **Columnar.** An aggregate of more of less parallel imperfect prismatic crystals, *e.g.*, amphibole.

(*v*) **Bladed.** An aggregate of flattened imperfect prismatic crystals, *e.g.*, kyanite.

(*vi*) **Fibrous.** An aggregate of fibres, which may or may not be separable, *e.g.*, asbestos.

(*vii*) **Foliated.** An aggregate of thin separable sheets ; also known as micaceous, *e.g.*, micas.

(*viii*) **Lamellar.** Mineral made up of separable plates or leaves which may be curved or straight, *e.g.*, gypsum.

(*ix*) **Tabular.** When the mineral shows broad flat surfaces as feldspar.

(*x*) **Granular or sacchroidal.** An aggregate of crystalline particles of about the same size, *e.g.*, chromite.

(*xi*) **Oolitic.** An aggregate of small spheres (fish scales).

(*xii*) **Pisolitic.** An aggregate of large spheres (shot like).

(*xiii*) **Botryoidal.** An aggregate like bunch of grapes.

(*xiv*) **Acicular.** An aggregate of needle-like crystals.

(*xv*) **Reniform.** Kidney-shaped aggregate.

(*xvi*) **Mammillary.** It consists of larger and mutually interfering prominences, *e.g.*, Malachite.

(*xvii*) **Stalactitic.** Cylindrical or conical form of minerals, generally due to deposition by dripping water.

2. Colour. It depends upon the absorption of some and reflection of others of the coloured rays which constitute white light. Some minerals show distinctive colours as follows.

(*a*) *White.* Calcite, Barite, Magnesite, Aragonite, Opal, Talc, Chalk etc.

(*b*) *Blue.* Azurite, Sodalite, Covellite, Lazulite, Lazurite, Apatite etc.

(*c*) *Green.* Fluorite, Beryl, Malachite, Microcline, Olivine, Epidote, Chlorite, Serpentine etc.

(*d*) *Yellow.* Sulphur, Marcasite, Chalcopyrite, Orpiment, Citrine (Quartz), Siderite etc.

(*e*) *Red.* Realgar, Jasper, Orthoclase, Pyrope, Zircon, Cinnabar etc.

(*f*) *Lead gray.* Galena, Graphite, Molybdenite etc.

(*g*) *Steel gray.* Hematite.

(*h*) *Brass yellow.* Pyrite.

(*i*) *Colourless.* Halite, Quartz, Calcite, Zeolite etc.

Variation in colour may be due to

(*i*) Surface alteration, (*ii*) difference in composition, (*iii*) presence of impurities and, (*iv*) inclusion of foreign matter.

3. Streak. It is the colour of the powder of a mineral in small amount and sometimes it is quite different from the colour of the mineral in mass. For example,

(*a*) siderite shows the streak colour as 'white',

(*b*) hematite shows the streak colour as 'cherry red',

(*c*) chalcopyrite shows black streak colour.

4. Lusture. It is the appearance of the surface of a mineral in reflected light. Lusture also depends upon absorption and reflection of light. Lusture of minerals differ both in intensity and kind depending upon the amount and manner of reflection respectively. Lusture may be of the following types.

(*a*) **Metallic lusture.** Gold, silver, copper, galena, graphite, molybdenite etc.

(*b*) **Non-metallic lusture :**

(*i*) *Vitreous lusture.* It is the lusture of a broken glass, *e.g.,* Quartz.

(*ii*) *Greasy lusture.* It is lusture of an oily glass, *e.g.,* Nepheline.

(*iii*) *Resinous lusture.* The lusture of resin as in sphalerite.

(*iv*) *Admantine lusture.* It is the lusture of a diamond.

(*v*) *Silky lusture.* It is shown by minerals possessing fibrous structure like asbestos, fibrous gypsum, fibrous calcite etc.

(*vi*) *Pearly lusture.* It is the lusture of a pearl, as in talc, opal, gypsum, kyanite etc.

(*vii*) *Earthy lusture.* It is a dull lusture as of kaoline, chalk.

When the degree of lusture is more, the surface shines like a mirror and it is known as splendent or brilliant lusture.

5. **Diaphaneity.** It is the degree of transparency of a mineral and is of the following types :

(*i*) *Transparent.* When almost all light falling on a mineral is transmitted through it.

(*ii*) *Semi-trans parent.* When the objects are seen through, but the outlines are not clear.

(*iii*) *Transluscent.* When light is transmitted through a mineral but the objects are not seen through it.

(*iv*) *Sub-transluscent.* When merely the edges are trans uscent.

(*v*) *Opaque.* When no light is transmitted through a mineral, it is said to be opaque.

6. **Phenomena depending on light.** Some minerals exhibit colours which are not in the minerals themselves but are produced by the effects of certain structures present in the minerals on white light. These are as

(*i*) *Play of colour.* A series of colours are seen at various angles like a rainbow.

(*ii*) *Change of colour.* Succession of colours produced when the mineral is turned about.

(*iii*) *Opalescence.* It is a pearly reflection from the interior of a mineral.

(*iv*) *Irridescence.* It is the brilliant colour shown by copper and pyrite.

(v) *Schillerization.* It is a kind of metallic colour shown by non-metallic hypersthene.

(vi) *Fluorescence.* Those minerals which after being exposed to ultra-violet light, emit light are known as fluorescent.

(vii) *Phosphorescence.* It is the property of continued emission of light after a substance has been subjected to rubbing, heating or electric radiation etc.

7. **Cleavage.** This is the property that some minerals exhibit of breaking along definite smooth planes. The presence of these planes is a simple indication of the difference in strength of bonds between atoms in the crystal : thus the property of cleavage is intimately connected with the atomic structure of minerals.

In number, there may be one or as many as six directions of cleavage. The atomic structure of minerals do not permit cleavage in five directions or in more than six directions.

Types of cleavages :

(i) **Pinacoidal.** Parallel to the pinacoidal faces. It is unidirectional :

(a) '*a*' *pinacoidal.* Kyanite.

(b) '*b*'-*pinacoidal.* Gypsum.

(c) '*c*'-*pinacoidal.* Also known as basal, cleavage, *e.g*, micas ,graphite, talc, etc.

In feldspar, there are '*b*-pinacoidal' and '*c*-pinacoidal' cleavages nearly at right angles.

(ii) **Prismatic.** Parallel to prism faces. It is bi-directional, which may or may not be at right-angles. Pyroxene and amphiboles show this type of cleavage.

(iii) **Cubic.** It is tri-directional at right angles as in galena, halite etc.

(iv) **Rhombohedral.** It is tri-directional, parallel to the faces of the rhombohedron as in the calcite group of minerals.

(v) **Octohedral.** Parallel to the faces of an octahedron. It is 4-direction as in case of diamond and fluorite.

(vi) **Pyramidal.** It is 4-direction, parallel to the pyramidal faces, *e g.*, scheelite.

(vii) **Dodecahedral.** Parallel to the faces of the dodecahedron, it is in 6-direction, as in sphalerite.

8. **Fracture :**

(i) **Conchoidal.** When a mineral breaks with curved concavities, more or less deep, as in a broken glass, *e.g.*, Quartz.

(*ii*) **Even.** When the fracture surface approximates to a plane.

(*iii*) **Uneven.** When the feacture surface is rough.

(*iv*) **Hackly.** When the surface is studded with jagged elevations and depression.

(*v*) **Splintery.** When the mineral separates out in fibres as in asbestos.

Fracture is, thus, the character of the surface obtained when a mineral is broken in a direction other than that of the cleavage.

9. Tenacity. It is the behaviour of a mineral under stress and it may be

(*i*) **Brittle.** When parts of mineral separates in powder, *e.g.*, calcite.

(*ii*) **Sectile.** When a mineral can be cut with a knife, but the slices yield under pressure, *e.g.*, Graphite, Gypsum etc.

(*iii*) **Malleable.** When the mineral can be cut with knife which flattens out under a hammer, *e.g.*, gold.

(*iv*) **Flexible.** When a mineral bends without breaking even when the force is removed, *e.g.*, micas, chlorite etc.

(*v*) **Elastic.** When the mineral attains its previous position after the withdrawl of the force, *e.g.*, mica.

10. Hardness. It is the resistance of a mineral offers to abrasion or scratching.

Moh's Scale of Hardness.

1. **Talc.**	2. **Gypsum.**
3. **Calcite.**	4. **Fluorite.**
5. **Apatite.**	6. **Feldspar.**
7. **Quartz.**	8. **Topaz.**
9. **Corundum.**	10. **Diamond.**

'Sclenometer' is an instrument used for determining hardness.

11. Specific gravity. It is the ratio of the weight for the mineral to the weight of an equal volume of water. It is determined through :

(*a*) *Walker's steel yard balance.* For large specimen.

(*b*) *Specific gravity bottle (Pycnometer).* For small mineral grains.

(*c*) *Chemical balance method.* For small fragments of minerals.

12. Magnetic property. A mineral capable of being attracted by a strong magnet is called magnetic, *e.g.*, magnetite and pyrrohite.

13. Electricity :

(a) *Pyroelectricity*. The development of positive and negetive charges of electricity on different parts of the same crystal when its temperature is suitably altered, is called pyroelectricity, *e.g.*, Quartz.

(b) *Piezo-electricity*. The property of development of electric-charges on crystallized mineral by pressure or by tension is called piezo-electricity, *e.g.*, Tourmaline, Quartz.

(c) *Photo-electricity*. When some minerals are exposed to radiation they produce electricity, *e.g.*, fluorite,

14. Radioactivity. Minerals containing elements of high atomic weights are called radioactive, because of their emissions. 'Greiger Counter' is an instrument used for the detection of radioactivity.

CHAPTER 33

PROPERTIES OF MINERALS

Optical properties of minerals, involve the various characteristics which can be observed through the petrological microscope and they include .

1. In Polarised Light :

(a) Form
— Euhedral, Perfect
— Subhedral
— Anhedral — Zig-zag outline

(b) Colour and pleochroism.

(c) Relief, whether positive or negative through Becke-line method.

(d) Cleavage and fracture.

(e) Inclusion.

2. Under Crossed-Nicols :

(i) Isotropism/Anisotropism.

(ii) Birefringence and interference colour.

(iii) Type of extinction.

(iv) Extinction angle, if any.

(v) Twinning and zoning.

(vi) Alteration, as seen by the microscope.

(vii) Association.

All these optical properties may be seen from the principles of optics described earlier.

Chemical Properties of Minerals :

Reaction with Acid. Excepting silicates and ferromagnesian minerals, usually, most of the carbonate-minerals and sulphide-minerals produces effervescence whenever they come in contact with acid.

The following chemical phenomena are usually found with minerals of various categories :

(*a*) **Isomorphism.** Chemical compounds which have an analogous composition and a closely related crystalline form are said to be isomorphous. The members of an isomorphous series, show a gradation in chemical composition, crystal forms, specific gravity, refractive-index, etc. from one extreme to the other. The plagioclase fieldspars constitute an excellent example of an isomorphous series.

(*b*) **Polymorphism.** This is the phenomenon in which substances containing the same chemical composition differs from one another by some physical properties like system of crystallization, hardness, density, etc. It is of the following types :

(*i*) On the basis of number of polymorphic forms :

(1) *Dimorphic.* When the chemical compound exists in two-distinct forms, e.g., calcite and aragonite, Pyrite and Marcasite, etc.

(2) *Trimorphic.* When the chemical compound exists in three distinct polymorphic forms, *e.g.,* Anadalusite, kyanite and sillimanite ; Rutile, anatase and brookite ; orthoclase, microcline and sanidine, etc.

The polymorphic forms in case of elements are known as allotrope.

(*ii*) **On the basis of reversibility :**

(1) *Enantiotropy.* When the polymorphic subtances are inter-changeable, it is called enantiotropy, for example, Diamond$\rightleftharpoons$Graphite and Quartz$\rightleftharpoons$Tridymite.

(2) *Monotropy.* When the changeability is from one substance to the other but not the reverse, *e.g.,* Marcasite$\rightleftharpoons$Pyrite.

Besides the above, it is to be noted that the high temperature polymorphs have higher degree of symmetry than its lower temperature polymorph which has lower degree of symmetry.

(*c*) **Pseudomorphism.** A mineral is said to exhibit pseudomorphism when it falsely assumes the outer form of a different mineral crystallizing in another system. It is due to : 1. Incrustation' 2. Infiltration, 3. Replacement, 4. Alteration, etc.

(*d*) **Homeomorphism.** Some minerals closely resemble others in crystal habits although they are different in chemical composition. If the geometry of the arrangement of the dissimilar ions is the similar appearing crystals may result, *e.g.,* Rutile and Zircon, which are tetragonal but have different chemical composition. Such minerals are called homeomorphs and the phenomenon as *homeomorphism.*

(*e*) **Paramorphism.** It is the phenomenon in which a crystal whose internal structure has changed to that of a polymorphous form without any change in the external form. Thus rutile changes to brookite and aragonite changes to calcite.

CHAPTER 34

STUDY OF COMMON ROCK-FORMING MINERALS

As we know, minerals are the individual units of rocks forming the crust of the earth. Thus rocks constitute the storehouse of minerals. On the basis of usefulness, all the minerals can be grouped into (i) Economic minerals and (ii) Rock-forming minerals.

The rock-forming minerals form a greater proportion of the rocks and are produced by normal mineral forming processes. Within the three major categories of rocks, the different rock-types are characterised by their own typical mineral assemblages. Silicates are by far the most important group of rock-forming minerals. Most of them and all the important ones can be grouped into 'groups' or 'families' having members of physical, chemical as well as optical similarities. Non-silicate minerals are much less important as rock-forming minerals. Of the 2000 known mineral species about 50 are rock-formers of which only about two dozens are most common.

The rock-forming minerals may be classified as follows :

1. Group-forming silicates⎫
2. Non-family silicates ⎬ Silicate Minerals.
3. Non-silicate minerals. ⎭

1 and 2. The silicate minerals are classified into various groups on the basis of the structural groups tney contain, i.e., according to their atomic-structure. Their fundamental unit in building of silicate-minerals is the SiO_4 group in which the silicon atom is situated at the centre of a tetrahedron, whose corners are occupied by four oxygen atoms. The various structures may be regarded as being derived from a tetrahedral unit (of SiO_4) by linking them together with the elimination of an oxygen atom at each linkage. There are mainly seven types of silicate structure, viz.

(a) **Neso-silicates.** The structure is that of independent (SiO_4)-tetrahedron. e.g., Olivine, Zircon, Garnet etc.

(b) **Soro-silicates.** Two tetrahedra sharing one oxygen, i.e., Si_2O_7, e.g., Melilite.

(*c*) **Cyclo-silicates.** These are closed rings of tetrahedra, sharing two oxygens, *i.e.*, the ratio between 'Si' and 'O' is 1 : 3. There are three types of rings also as

(*i*) each of 3-tetrahedra sharing an oxygen atom, *e.g*, Benitoite,

(*ii*) each of 4-tetrahedra sharing an oxygen atom, *e.g.*, Axinite.

(*iii*) each of 6-tetrahedra sharing an oxygen atom, *e.g.*, Beryl.

(*d*) **Ino-silicates.** These are also known as chain structure and are of two types :

(*i*) *Single-chain structure.* In this case two oxygen atoms of the silicon-tetrahedron are shared. Si : O=1 : 3. These constitute continuous single chains of tetrahedra, *e.g.*, Pyroxene.

(*ii*) *Double-chain structure.* These are continuous double chains of tetrahedra alternately sharing two and three oxygen. The ratio of silicon to oxygen is 4 : 11, *e.g.*, Amphibole. These are also known as *Band structure.*

(*e*) **Phyllosilicates.** These are continuous sheets of tetrahedra sharing three oxygens. Ratio of silicon to oxygen is 4 : 10, *e.g.*, Mica. Also known as sheet structure.

(*f*) **Tekto silicates.** These are three-dimensional frame-work of tetrahedra with all four oxygen atoms shared. Also known as Frame work structure. Ratio of silicon to oxygen is 1 : 2, *e.g.*, Quartz, Felspar etc.

Accordingly the following major silicate families have been recognised :

(*a*) Olivine group.	(*b*) Silica group.
(*c*) Felspathoid group.	(*d*) Pyroxene group.
(*e*) Amphibole group.	(*f*) Mica group.

Other silicate groups which are less abundant, are

1. Garnet group.	2. Epidote group.
3. Zeolite group.	4. Kaoline group.

5. Alumino-silicates.

The other silicate minerals which do not form groups are as follows.

(*i*) Zircon, (*ii*) Sphene, (*iii*) Topaz, (*iv*) Staurolite, (*v*) Beryl, (*vi*) Cordierite, (*vii*) Tourmaline, (*viii*) Talc, (*ix*) Serpentine. x) Dumortierite etc.

3. Non-silicate minerals. These are oxides, hydroxides, carbonates, sulphides, sulphates etc. Among the non-silicates the oxides are most important since many of them occur as minor accessories of rocks.

Minerals like quartz, feldspar, mica, amphiboles, pyroxenes and olivines are the common constituents of igneous rocks of different igneous types.

In case of metamorphic rocks minerals like kyanite, sillimanite, staurolite, andalusite, chlorite, garnet etc. are commonly found.

As the sedimentary rock are formed due to the consolidation of the weathering products of the pre-existing rocks, they may contain any mineral or assemblage of minerals. Study of the specific minerals which are commonly treated as rock-forming minerals are dealt separately.

CHAPTER 35

SILICATE STRUCTURE

The class of silicate minerals is of greater importance than any other, for about 25% of the known minerals and nearly 40% of the common ones are 'silicates'. The silicates make up 90% of the earth's crust. Of every 100 atoms in the crust of the earth, more than 46 are oxygen, over 27 are silicon and 7 to 8 are aluminium. The crust has been pictured as a box-work of oxygen-ions bound together by the small highly charged silicon and aluminium ions. The interstices of this more or less continuous silicon-oxygen-aluminium net-work are occupied by ions of magnesium, iron, calcium, sodium and potassium. The predominance of alumino-silicates and silicates reflects the abundance of oxygen, silicon and aluminium.

All the silicate structures, so far investigated show that the silicon atoms are in four-fold co-ordination with oxygen. This arrangement appears to be universal in these compounds and the bonds between silicon and oxygen are so strong that the four oxygen are always found at the corners of a tetrahedron of nearly constant dimensions, and regular shape, whatever the rest of the structure may be like.

The radius ratio of the four valent silicon ion to that of oxygen ion is equal to 0.318 [since Si (radius)$=0.42$Å and oxygen (radius)$=1.32$Å]. It indicates that four-fold co-ordination will be the stable state for silicon-oxygen grouping. Although electron-sharing is present in the silicon-oxygen bond, the total bonding energy of the silicon ion is still disturbed equally among its four closest oxygen neighbours. Hence, the strength of any single silicon-oxygen bond is equal to just one-half the total bonding energy available in the oxygen ion. Each oxygen ion has, therefore, the potentiality of bonding to another silicon ion and entering into another tetrahedral grouping, thus uniting the tetrahedral groups through the shared oxygen. This sharing may involve one, two, three or all four of the oxygen ions in the tetrahedron, giving rise to a diversity of structural configurations.

Sharing of one oxygen between any two adjacent tetrahedra may, if all oxygens are so shared give rise to structures with a very high degree of connectivity, such as quartz structure. This linking of tetrahedron by sharing of oxygen is known as polymerisation. It has been observed that the higher the temperature of formation the lower the degree of polymerisation and *vice-versa*. It has long been noted that the silicate minerals in igneous rocks display a fairly regular and predictoble sequence of crystallization, beginning with olivine and progressing through pyroxene to amphibole and thence to mica. This sequence is in the order of increasing polymerization of the silicate tetrahedra.

Depending on the degree of polymerisation and the extent of oxygen-sharing between tetrahedra, the silicate frame-work may consist of separate tetrahedra, separate multiple tetrahedral groups, chains, double chains, sheets or three-dimensional box-works.

Up to the 1930s, the analyses of silicates were interpreted and their formulas generally written in terms of a number of hypothetical oxy-acids of silicon. Thus, olivine Mg_2SiO_4, was termed an ortho-silicate and considered to be a salt of orthosilicic acid H_4SiO_4 ; Enstatite, $MgSiO_3$ was called a metasilicate and considered to be a salt of metasilicic acid H_2SiO_3. The silicate structures, so far recognized are of the following types :

1. Neso-silicates. These are independent or isolated SiO_4-tetrahedra which are bound to each other only by ionic bonds through interstitial cations. Their structures depend chiefly on the size and charge of the interstitial cations.

Considering the valencies of the elements composing the SiO_4 group, it is found that silicon has four positive and each oxygen has two negative valencies. Thus there are eight negative valencies in all and the group as a whole therefore has four negative valencies in excess. In olivine structure, cations (mainly Mg) lie between the tetrahedral groups and contribute the necessary '+ve' charges to make the structure electrically neutral.

Minerals. Olivine, Garnet, Zircon, Sillimanite, Kyanite, Andalu-site, Staurolite, Phenacite, Topaz, Willemite, Sphene etc.

2. Soro-silicates. The soro-silicates are characterised by isolated double-tetrahedral groups formed by two SiO_4 tetrahedra sharin g a single apical oxygen. The resulting ratio of silicon to oxygen ^{1}s 2 : 7. They have a net charge of '—6'. As the charge is '—6', three divalent ions are needed to balance it.

Minerals. Idocrase, minerals of epidote group. Melilite $(Ca_2MaSi_2O_7)$, Lawsonite $[CaAl_2(Si_2O_7)(OH)_2H_1O)]$, Hemimorphite $[Zn_4Si_2O_7(OH)_2.H_2O]$ etc.

3. Cyclosilicates. When each SiO_4-tetrahedron shares two of its oxygen with neighbouring tetrahedra, they may be linked into rings. They have a ratio of $Si : O = 1 : 3$. Three possible closed cyclic configurations of this kind may exist as (*a*) Each of the three tetrahedra shares an oxygen atom.

(*b*) Each of the four tetrahedra shares an oxygen atom.

(*c*) Each of the six-tetrahedra shares an oxygen atom.

These are all having the formula which are multiples of SiO_3. The simplest is the Si_3O_9 ring represented among minerals only by the rare titanosilicate, Benitoite (hexagonal)-$BaTiSi_3O_9$.

The Si_4O_{12} ring occurs together with BO_3-triangles and (OH) groups in complex structure of the triclinic mineral '*axinite*'.

The Si_6O_{18} ring, however is the basic frame work of the common important minerals like beryl and tourmaline. The hexagonal Si_6O_{18} rings are arranged in planar sheets parallel to (0001) in beryl. These sheets are so firmly bonded by the small beryllium and aluminium ions with their high surface density of charge and high polarizing power that only poor cleavage results.

In tourmaline, however, the rings are polar. This polarity of the fundamental structural unit leads to the well known polar character of the tourmaline crystal.

The cyclosilicates are also known as 'Ring-structures'.

4. Chain structures. These are also known as '*ino-silicates*'. Here SiO_4-tetrahedra are joined together to form chains of indefinite extent.

There are two-principal modifications of this structure yielding somewhat different composition :

(*a*) Single chains, in which Si : O is 1 : 3 characterised by the pyroxenes and pyroxenoids.

(*b*) Double chains, where alternate tetrahedra in two parallel single chains are cross-linked and the Si : O ratio is 4 : 11, characterised by the amphiboles.

(*a*) **Single-chain Structures.** The chains consist of a large number of linked SiO_4 groups, each sharing two oxygens and have the composition $n(Si_2O_6)$. Here the excess negative charge on the Si_2O_6 chain is balanced by the valencies of other cations. The chains run parallel to the '*c*-axis' of the mineral and are bonded together by the calcium and magnesium ions which lie between them.

(*b*) **Double-chain structure.** These are also known as the 'Band-structures'. Here the alternate tetrahedra are arranged in two parallel ways and these chains are indefinite in extension and elongated usually in '*c*'-crystallographic direction' and are bound by metallic ions.

Sheet structures (Si_4O_{10}). It is also known as '*phyllosilicates*'. It is formed when the SiO_4 tetrahedra are linked by three of their corners and extend indefinitely in a two-dimensional net-work or sheet, which has a silicon and oxygen ratio of 4 : 10. This is the fundamental unit-in all mica and clay-structure.

The sheets form a planar net-work responsible for the principal characteristics of minerals of this type—their pronounced pseudo-hexagonal habit and perfect basal cleavage parallel to the plane of the sheet. Most of the minerals of this class are hydroxyl bearing. Most of the members have platy or flaky habit and one prominent cleavage. They are generally soft, of relative low specific gravity and may show flexibility or even elasticity of the cleavage lamellae.

Depending on the mode of co-ordination of the hydroxyl

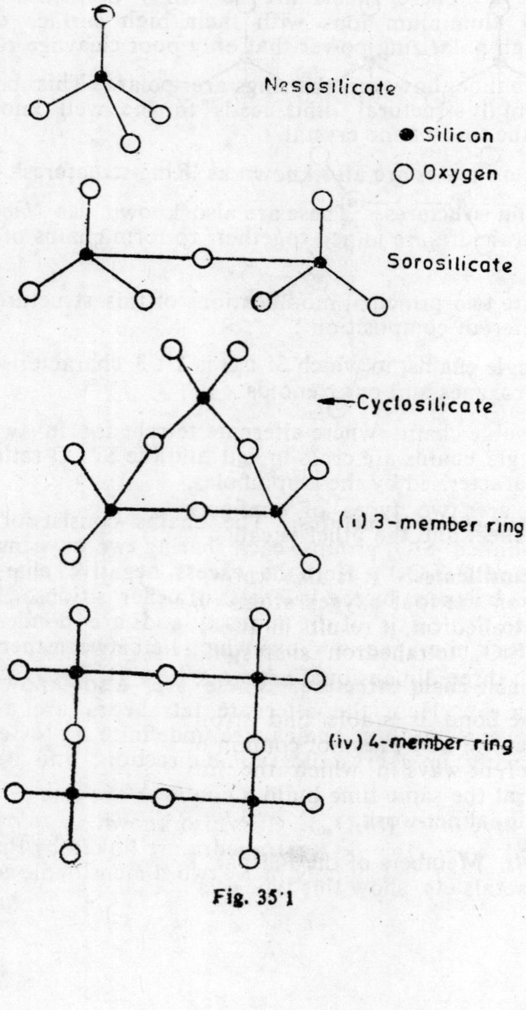

Fig. 35·1

(iii) 6-member ring

(i) Single chain

(ii) Double chain

Chain structure

Fig. 35'2

group there are two types of configuration; one is called the dioctahedral sheet and the other is called the trioctahedral sheet.

6. **Tectosilicates.** It is also known as *'framework structure'*. When each of the four oxygen atoms of each tetrahedron is shared by another tetrahedron, it results in the formation of tectosilicates. Here every SiO_4 tetrahedron shares all its corners with other tetrahedra giving a three-dimensional net-work in which Si : O=1 : 2.

Here the bond is stable and strong and the frame-work is electrically neutral and does not contain other structural unit. There are eight different ways in which the linked tetrahedra may share oxygen and at the same time build a continuous electrically neutral three dimensional net-work.

Minerals. Members of the Felspar, Felspathoid, Zeolite, Quartz group of minerals etc. show this type of silicate structures.

CHAPTER 36

OLIVINE

This is a group of rock-forming silicates. The minerals are olive-green or brown in colour. All these minerals crystallize in orthorhombic system.

1. Atomic structure. These are neso-silicates which essentially consists of a series of isolated SiO_4 tetrahedra which are linked by means of metal cations.

2. Chemical composition. The group includes minerals which may be represented by the formula R_2SiO_4, where $R = Mg$ or Fe. The members of this group belong to a continuous series of solid-solutions between Forsterite Mg_2SiO_4 and Fayalite Fe_2SiO_4. Common olivine is intermediate between them with excess Mg and the formula is represented as $(Mg, Fe)_2SiO_4$.

Various members are there in the isomorphous series, in which the end members are forsterite and fayalite, *viz.*

Forsterite	$Fo_{100\%}$	$Fa_{0 \text{ to } 10\%}$
Chrysolite	$Fo_{90 \text{ to } 70\%}$	$Fa_{10 \text{ to } 30\%}$
Hyalosiderite	$Fo_{70 \text{ to } 50\%}$	$Fa_{30 \text{ to } 50\%}$
Hortonolite	$Fo_{50 \text{ to } 30\%}$	$Fa_{50 \text{ to } 70\%}$
Ferrorhortonolite	$Fo_{30 \text{ to } 10\%}$	$Fa_{70 \text{ to } 90\%}$
Fayalite	$Fo_{10 \text{ to } 0\%}$	$Fa_{90 \text{ to } 100\%}$

Other allied minerals :

Knebelite $FeMn_2SiO_4$	Monticellite $CaMgSiO_4$
Larsenite $PbZnSio_4$	Tephorite Mn_2SiO_4,

3. Physical properties :

1. Crystal system	Orthorhombic
2. Colour	Olive green
3. Streak	Colourless
4. Lusture	Vitreous lusture

5.	Hardness	6·5 to 7
6.	Cleavage	Absent
7.	Sp. gravity	3·2 to 4·3
8.	Fracture	Conchoidal
9.	Twinning	Rare.

4. Optical properties :

(*i*) Colourless and non-pleochroic.

(*ii*) *Ref. index.* High positive relief, High Ref. index.

(*iii*) *Bi-refringence.* Strong 0·037.

(*iv*) *Extinction.* Straight.

5. Varieties. The gem-quality of olivines are referred to as peridot.

6. Alterations. Olivine commonly alters to serpentine (antigorite) and secondary iron-oxide. In basaltic rocks, the alteration of the outer iron-rich rim of olivine produce a brownish red mineral called 'Iddingsite'.

7. Occurrence. It characterises the ultra-basic igneous rocks as dunites, peridotites serpentinites and basic rocks like norite, gabbro, dolerite, basalt, etc. The common associates are chromite, spinel, pyrope etc. Olivine and quartz (primary) never occur together.

Forsterite is formed by dedolomitisation or contact metamorphism of magnesium rich sedimentary rocks as dolomitic limestone rich in silica.

CHAPTER 37

PYROXENES

These rocks forming silicates contain the Si_2O_6 single chain structure (inosilicates). These are anhydrous silicates of Mg and Fe and thus are predominantly found in ferro-magnesian rocks, i.e., in basic and ultrabasic rocks. The following minerals have been included in the pyroxene group :

1. Orthorhombic :

   (a) Enstatite $MgSiO_3$.

   (b) Bronzite   ('Fe' rich enstatite).

   (c) Hypersthene   Mg, $FeSiO_3$.

   (d) Ferrohypersthene $FeMgSiO_3$.

   (e) Eulite   Mg content 10 to 30%.

   (f) Ferrosilite $FeSiO_3$.

(A) 2. Monoclinic :

   (a) Clino-enstatite $Mg_2Si_2O_6$.

   (b) Diopside $CaMgSi_2O_6$.

   (c) Sahlite $Ca(Mg, Fe)Si_2O_6$.

   (d) Hedenbergite $CaFeSi_2O_6$.

   (e) Augite $(Ca, Mg, Fe, Al)_2AlSi_2O_6$.

   (f) Pigeonite $(CaMg)(MgFe)Si_2O_6$.

(B) Alkali-pyroxenes :

   (a) Aegirine $NaFeSi_2O_6$.

   (b) Jadeite $NaAlSi_2O_6$.

   (c) Spodumene $LiAlSi_2O_6$.

   (d) Johannsenite $CaMn, Si_2O_6$.

3. Triclinic. They are not related structurally to the pyroxenes, although chemically they have identical formulae. They have a single chain of linked SiO_4 tetrahedra, which is not the simple chain of the pyroxenes. They are commonly known as "pyroxenoids", and include minerals like Wollastonite ($CaSiO_3$), Pectolite [$Ca_2NaH(SiO)_3$] and Rhodonite ($MnSiO_3$), Bustamite [$MnCa(SiO_3)_2$].

Physical Properties of Pyroxenes. These are usually prismatic crystals.

1. *Colour.* Nearly black or green of various shades.

2. *Lustre.* Vitreous to subvitreous. Hypersthene shows a kind of metallic-pearly lusture termed 'Schillerisation'.

3. *Cleavage.* 2 sets, prismatic at angles 87° and 93°.

4. *Hardness.* 5 to 6.

5. *Sp. gravity.* Low to moderate.

6. *Twinning.* Contact twins in case of monoclinic members.

Optical Characteristics

(*a*) **Orthopyroxenes :**

(*i*) Green and pleochroic (except in enstatites).

(*ii*) *Ref. index.* High. Optically +ve.

(*iii*) *Interference colour.* 1st order.

(*iv*) *Extinction.* Parallel.

(*b*) **Clinopyroxenes :**

(*i*) Colourless to pale green and pleochroic.

(*ii*) *Ref. index.* Higher than Canada Balsam.

(*iii*) *Int. colour.* 2nd order.

(*iv*) *Extinction.* Inclined, 45°.

Varieties :

(*i*) *Diallage.* Translucent and is fibrous augite.

(*ii*) *Omphacite.* Foliated diopside, found in Eclogite.

(*iii*) *Kunzite.* Gem-variety of spodumene.

(*iv*) *Hiddenite.* Emrald green spodumene (gem-variety).

Occurrence. Mostly found in basic rocks like Gabbro and their hypabbyssal and volcanic equivalent,ultrabasic rocks like—peridotite, pyroxenite also contain predominantly pyroxenes. Hypersthene is characteristic of norites and charnockites. Spodumen occurs as large crystals in pegmatites.

CHAPTER 38

AMPHIOBOLES

These are hydrous ferro-magnesian silicate minerals and along with pyroxene constitutes about 1/5th of all the known rocks.

Atomic structure. Inosilicates characterised by double chain structure.

Chemical composition. Hydrous silicates of Magnesium, Iron, Calcium, Aluminium as well as Alkali metals.

Physical properties :

1. Form. Some minerals of the group are distinctly orthorhombic, and monoclinic, others are triclinic in their crystallisation. [Accordingly each group shows different characteristics under microscope].

(*a*) *Orthorhombic.* (*i*) Anthophyllite $(MgFe)_7(Si_4O_{11})(OH)_2$.

(*b*) *Monoclinic :*

(*i*) Cummingtonite	(*ii*) Grunerite
(*iii*) Tremolite	(*iv*) Actinolite
(*v*) Hornblende	(*vi*) Edenite
(*vii*) Tschermakite	(*viii*) Pargasite
(*ix*) Hastingsite	(*x*) Kaersutite
(*xi*) Lamprobolite	(*xii*) Arfvedsonite
(*xiii*) Glaucophane	(*xiv*) Riebeckite.

(*c*) *Triclinic.* Cossyrite.

2. Colour. Most of the members are greenish black in colour.

3. Lusture. Vitreous.

4. Cleavage. 2 sets, perfect, prismatic (110), at 56° and 124°.

5. Hardness. 5 to 6.

6. **Sp. gravity.** 2·5 to 3·5.

7. **Twinning.** Contact twins are common.

Optical properties :

(*i*) Green colour and commonly pleochroic.

(*ii*) *Ref-index.* High.

(*iii*) *Interference colour.* $\left\{\begin{matrix}\text{Orthoamphiboles}\\\text{Clinoamhiboles}\end{matrix}\right\}$2nd order.

(*iv*) *Extinction.* Orthoamphiboles parallel.
Clinoamphiboles 10° to 20°.

Varieties :

(*i*) *Smargdite.* Foliated act‍ionolite.

(*ii*) *Uralite.* Secondary tremolite-actinolite.

(*iii*) *Asbestos* :

    (*a*) Amosite—Fibrous anthophyllite.

    (*b*) Actinolite—Asbestos proper.

    (*c*) Crocidolite—Fibrous riebeckite.

    (*d*) Chrysotile—Fibrous serpentine.

(*iv*) *Nephrite.* Fine grained tremolite-actinolite.

Occurrence. Mostly occur in metamorphic rocks and also in igneous rocks. Crocidolite occurs in highly siliceous metamorphic rocks.

CHAPTER 39

FELDSPARS

This is the most important group of the rock-forming silicates. They constitute about 2/3rd of the igneous rocks.

Atomic structure Tektosilicates :

Chemical composition. They are alumino-silicates of potassium, sodium, calcium and barium, and may be regarded as isomorphous mixtures of the four-end members given below, of which the first three are common while the fourth is rare.

1. Orthoclase and Microcline $KAlSi_3O_8$.
2. Albite $NaAlSi_3O_8$.
3. Anorthite $CaAl_2Si_2O_8$.
4. Celsian $BaAl_2Si_2O_8$.

The three isomorphous series are :

(a) **Anorthoclase series.** Between a-'K-and a-'Na' feldspar.

(b) **Hyalophane series.** Between a-'K' feldspar and a-'Ba'-feldspar with K > Na.

(c) **Plagioclase series.** Between albite and anorthite :

(i) Albite   (ii) Oligoclase

(iii) Andesine   (iv) Labradorite

(v) Bytownite   (vi) Anorthite.

Physical Properties :

(i) **Crystal form.** Orthoclase is monoclinic. Microline and other plagioclases are triclinic.

(ii) **Colour.** Orthoclase is flesh-red in colour. Microcline is green in colour and the colour of plagioclases ranges from white to gray.

(iii) **Lusture.** Vitreous or pearly (play of colour is marked).

(iv) **Cleavage.** 2 sets—one parallel to (C01) face and other to

(010). The angle between the cleavages is 90° in case of orthoclase but less than 90° in other members.

(*v*) **Hardness.** 6 (six).

(*vi*) **Sp. gravity.** 2·5 to 3, according to calcium content.

(*vii*) **Twinning :**

(*a*) (1) *Carlsbad.*—'*c*' axis is the twin axis and (010) the composition plane.

(2) *Baveno.* (021) is the twin plane (*i e.*, clinodome).

(3) *Manebach.* (001) is the twin plane.

These three are commonly found in orthoclase.

(*b*) **Microcline :**

(1) *Albite.* Twin plane (001), twin axis perpendicular to this.

(2) *Pericline law.* (010) composition plane, twin axis is the *b*-axis.

(*c*) **Plagioclases :**

They show all the above twinning types.

Microscopic characteristics :

1. *Form.* Subhedral to anhedral.

2. *Cleavage.* 2 sets.

3. *Bi refringence.* Weak (Ref. index=low), *i.e.*, 0·009.

4. *Twinning* :

(*i*) Orthoclase : simple twinning.

(*ii*) Microcline : cross-hatched twinning.

(*iii*) Plagioclase : polysynthetic (or lamellar).

5. *Extinction angle.* 15 to 30°. Michael-Levy method is used to determine this.

Varieties :

(1) *Sanidine.* A high temperature potassium feldspar (in volcanic rocks).

(2) *Adularia.* Low-temperature orthoclase.

(3) *Moon stone.* Opalescent adularia or albite.

(4) *Aventurine.* Gem variety of albite.

(5) *Amazon stone.* Bright green microcline.

Occurrence. Alkali-feldspars, *i.e.*, orthoclase, microcline and albite are common in acid igneous rocks like granite, grano-diorite, syenite etc. in sandstones and in gneisses (metamorphic rocks).

Plagioclase occurs in intermediate to basic igneous rocks.

Important Features of Occurrence :

1. **Perthite.** Intergrowth between orthoclase/microcline and albite or obligoclase. But albite patches found in microcline mass.

2. **Antiperthite.** Microcline is found crystallised in predominant albite (as ground mass).

3. **Graphic.** Intergrowth between quartz and orthoclase feldspar.

4. **Myrmekitic.** Intergrowth between quartz and plagioclase feldspar.

5. **Zoning.** Variation of composition marked in the marginal region of feldspar-minerals in zones.

CHAPTER 40

QUARTZ

Quartz is a member of the silica group of minerals, which have tekto-silicate structure and the chemical composition is SiO_2. The atomic structure as is found in the crystalline varieties does not apply to silica-glass (*i.e.*, Lechatelierite). The *crystalline* variety occurs in the following distinct forms :

(*i*) Quartz   (*ii*) Tridymite

(*iii*) Cristobalite   (*iv*) Coesite

(*v*) Stishovite   (*vi*) Keatite.

There are the three crystalline polymorphs of silica. While quartz is a low-temperature polymorph formed below 870°C, trydimite is formed between 870° to 1470°C and cristobalite is formed at a temperature above 1470°C.

The non-crystalline variety of silica occurs as

(*i*) Lechatelierite   (*ii*) Opal, and

(*iii*) Chalcedony.

Physical characteristics :

1. **Form.** White quartz is a member of the Hexagonal system, Tridymite belongs to Orthorhombic system and Cristobalite belongs to Isometric system.

2. **Streak.** White.

3. **Lusture.** Vitreous to sub-vitreous.

4. **Hardness.** 7 (seven).

5. **Cleavage.** No cleavage (an important characteristic).

6. **Sp. gravity.** Low, *i.e.*, 2·65.

7. **Twinning** (*i*) Common twins are *Dauphine type*, a penetration twin with the '*c*'-axis as the twin axis.

(*ii*) *Brazil type*. A penetration twin with (11$\bar{2}$0) as the twin plane.

(*iii*) *Japanese law*. Contact twins with (11$\bar{2}$2) as twin plane.

8. **Electric-property.** Quartz is 'piezo' as well as 'pyro'-electric.

9. **Colour**. Quartz is colourless but the non-crystalline varieties are coloured.

Optical properties :

(*i*) Non-pleochroic colourless.

(*ii*) *Ref. index.* Low $+ve$, Birefringence$= 0\cdot009$.

(*iii*) *Polarisation colour.* 1st order gray or yellow.

(*iv*) *Extinction.* Wavy (Glassy varieties are isotropic). In Brazil-twinned plates-'*Airy's spiral*' is seen.

Polymorphs of quartz. Besides tridymite, cristobalite, ccesite, stishovite, there are two modifications of quartz, *viz.*, α-quartz and β-quartz.

(*a*) α-**quartz.** Low-temperature quartz, formed below 573°C.

(*b*) β-**quartz.** High-temperature quartz, formed between 573° and 870°C (found in volcanic rocks as phenocrysts).

Varieties of quartz basing on colour :

(*i*) *Rock-crystal.* Clear and transparent.

(*ii*) *Amethyst.* Violet.

(*iii*) *Citrin.* Pale yellow.

(*iv*) *Morion.* Black quartz.

(*v*) *Cat's eye.* Fibrous variety, showing opalescence due to the presence of titanium.

(*vi*) *Rose quartz.* Pink.

(*vii*) *Milky quartz.* White.

Non-crystalline variety. Excepting silica glass 'Lechatelierite', the other two, *i.e.*, opal and chalcedony contains some water $SiO_2.nH_2O$.

Opal is of some varieties like Fire opal, siliceous sinter or geyserite, diatomite, etc.

Chalcedony is having varieties like agate, jasper, sard, chert, flint, chrysoprase, blood stone (*i.e.*, carnelian), horn stone etc.

Occurrence. It is present in silica rich igneous rock. It is the basic material of sandstone and is found in metamorphic rocks like gneisses, schists, charnockites and khondalites.

CHAPTER 41

MICA

Micas constitute an important group of rock forming minerals, as ferromagnesian silicates, next in importance to the amphiboles and pyroxenes. They form a link between felspar and felspathoids *i.e.*, the light coloured constituents of the igneous rocks and the dark coloured minerals.

1. **Atomic structure.** These are phyllosilicates which possess the Si_4O_{10} sheet structure.

2. **Chemical composition.** These are hydrated alumino silicates of K, Na or Li, with Mg or Fe in darker members.

3. **Classification.** Some classification of mica groups are as

(*a*) *Dioctahedral group.* Muscovite.

(*b*) *Trioctahedral group.* Phlogopite and Biotite.

According to other classifications, the composition of the micas are taken into account and they are as

(I) **Muscovite group.** These are aluminosilicates of alkali metals without Mg or Fe ; and are colourless. It includes members like :

(*i*) *Muscovite.* K-aluminosilicate with (OH).

(*ii*) *Paragonite.* Na-aluminosilicate with (OH).

(*iii*) *Lepidolite.* K-Li-Al, silicate with (OH).

(II) **Biotite group.** In addition to the alkali metals, they also contain Mg, Fe and are dark in colour. They contain the following members :

(*i*) *Biotite.* K, Mg, Fe, Al, silicate with (OH).

(*ii*) *Phlogopite.* K, Mg, Al, silicate with (OH).

(*iii*) *Zinnwaldite.* Li, Fe-Al, silicate with (OH).

4. **Physical properties :**

(*a*) They crystallise in monoclinic system but the forms are pseudo-hexagonal.

(b) *Moscovite.* Colourless.

(c) *Biotite.* Dark coloured.

(d) *Lusture.* Pearly, splendaht.

(e) *Cleavage.* Perfect basal cleavage.

(f) *Hardness.* 2 to 3.

(g) *Sp. gravity.* Low.

(h) *Twinning.* Rarely seen.

(i) *Special characteristics.* They possess unique combination of properties of elasticity, toughness, flexibility, transparency, resistance to heat and property of splittings into thin films.

5. **Optical properties :**

(i) Muscovites colourless to pale green, Biotites dark brown.

(ii) Coloured varieties are pleochroic.

(iii) *Ref. index.* High.

(iv) *Bi-refringence.* Strong, i.e., 0.033 to 0.059.

(v) *Extinction.* Parallel.

6 **Varieties :**

(i) *Sericite.* Secondary mica (muscovite).

(ii) *Fuschsite.* Cr-bearing mica.

(iii) *Gilbertite.* Fl-rich muscovite.

(iv) *Roscoelite.* Vanadium bearing muscovite.

(v) *Damourite.* Secondary muscovite by hydrothermal alteration of feldspar.

(vi) *Lepidomelane.* An iron-rich biotite.

(vii) *Vermiculites.* Altered biotite.

7. **Occurrence.** Muscovite occurs in acid-igneous rocks. Biotite occurs in all igneous rocks, particularly the acid and intermediate ones.

In sedimentary rocks also they are present as an authigenic constituent.

In schists and gneisses (metamorphic rocks) micas are the usual constituents.

Important characteristics. Micas show a peculier optical property known as '*birds eye structure*', which is noted as the extinction is approached.

Percussion figures are also the characteristic features found in mica, when any pointed object is pressed against it.

CHAPTER 42

CHLORITES

These are micaceous minerals which are green in colour and have cleavage flakes which are flexible but not elastic. They are devoid of alkali metals and rich in ferrous iron. They show a type of sheet structure and are hydrous. The important species of chlorite groups are as follows :

    1. Clinochlore. 2. Penninite. 3. Prochlorite.

Physical properties :

  (*i*) They all crystallise in the monoclinic system.

  (*ii*) *Cleavage.* Perfect.

  (*iii*) *Hardness.* 1 to 2·5.

  (*iv*) *Sp. gravity.* 2·5 to 3.

  (*v*) *Twinning.* Not recognisable in handspecimen.

Optical properties :

  (*a*) Green in colour and pleochroic.

  (*b*) *Ref. index.* Higher than Canada Balsam.

  (*c*) *Birefringence.* Very weak = 0·001 to 0·011.

  (*d*) *Interference colour.* First order.

  (*e*) *Extinction.* Parallel.

Variety :

    1. *Chamosite.* An iron-rich chlorite.

    2. *Chlorophaeite.* A chlorite formed by the alteration of volcanic glass.

    Occurrence. These are always secondary resulting by the metamorphism or hydrothermal alteration of ferromagnesian minerals.

CHAPTER 43

GARNETS

This is a group of six minerals which are isomorphous in nature, and never occur pure as represented by their chemical composition.

1. Atomic structure. These are neso-silicates, formed of independent SiO_4 tetrahedra.

2. Chemical composition. The general formula for the garnets is $R_3{}^{11}R_2{}^{111}Si_3O_{12}$, where R^{11} can be Fe^{11}, Mg, Mn^{11} or Ca, and R^{111} can be Fe^{111}, Al or Cr.

3. Classification. Mainly there are two groups as :

(*i*) **Pyralspite.** Containing three members like :
- (*a*) *Pyrope.* $Mg_3Al_2(SiO_4)_3$.
- (*b*) *Almandite.* $Fe_3Al_2(SiO_4)_3$.
- (*c*) *Spessartite.* $Mn_3Al_2(SiO_4)_3$.

(*ii*) **Ugrandite.** It consisting of three members like :
- (*a*) *Uvarovite.* $Ca_3Cr_2(SiO_4)_3$.
- (*b*) *Grossularite.* $Ca_3Al_2(SiO_4)_3$.
- (*c*) *Andradite.* $Ca_3Fe_2(SiO_4)_3$.

4. Physical properties :

(*i*) Garnets are externally cubic in form (Rhombododecahedron).

(*ii*) Any colour except blue ; grossularite, uvarovite green, pyrope, almandite, and spessartite are deep red to black in colour, while andradite is black in colour.

(*iii*) *Lusture.* Vitreous to sub-vitreous.

(*iv*) *Cleavage.* None.

(*v*) *Hardness.* 7.

(*vi*) *Sp. gravity.* High.

(*vii*) *Twinning.* None.

5. Optical properties :

(*i*) Non-pleochroic colourless, sometimes pale pink in colour. Uvarovite is green in colour.

(*ii*) *Ref. index.* Very high.

(*iii*) *Under crossed nicol.* Dark, *i.e.*, isotropic.

6. Varieties :

(*i*) *Melanite.* Black andradite.

(*ii*) *Schorlomite.* Titaniferous andradite.

(*iii*) *Demontoid.* Clear-green andradite.

(*iv*) *Hessonite.* Also known as cinnamon stone is a yellow-brownish variety of grossularite.

(*v*) *Topozolite.* Yellow-andradite.

Occurrence. Garnets are usually metamorphic minerals, but some also occur in igneous and sedimentary rocks. But everywhere they are characterised by their typical rhombododecahedron and trapezohedron forms.

CHAPTER 44

CARBONATES

The carbonate minerals do not form a homogeneous group, while some are hexagonal in crystallisation, others are orthorhombic crystals and still some others are massive in nature.

1. Hexagonal carbonates :

The more important members are :

(*i*) *Calcite.* $CaCO_3$.　　　　(*ii*) *Dolomite.* $CaMg(CO_3)_2$.

(*iii*) *Magnesite.* $MgCO_3$.　　(*iv*) *Siderite.* $FeCO_3$.

(*v*) *Rhodochrosite.* $MnCO_3$.　(*vi*) *Smithsonite.* $ZnCO_3$.

(*vii*) *Ankerite.* $(Ca, Mg, Fe)CO_3$.

Physical properties :

(*i*) All are rhombohedral crystals.

(*ii*) *Colour.* Calcite is colourless, dolomite is often tinged with yellow, Magnesite is white or colourless, Siderite is yellowish brown, Rhodochrosite is pink, white-green colour is found in Smithsonite and in case of Ankerite.

(*iii*) *Lusture.* Vitreous to silky in fibrous varieties.

(*iv*) *Cleavage.* Perfect, rhombohedral (3 sets).

(*v*) *Hardness.* Calcite-3, in others it is 3˙5 to 5.

(*vi*) *Sp. gravity.* Low.

Optical properties :

(*i*) Non-pleochroic colourless.

(*ii*) *Ref. index.* Low.

(*iii*) *Bi-refringence.* High, in calcite it is 0˙18.

(*iv*) *Twinning.* Polysynthetic.

(*v*) Twinkling is usually observed.

(*vi*) *Extinction.* Depends on the orientation of the section.

Varieties :

(*i*) *Iceland spar.* Transparent ealcite.

(*ii*) *Satin spar.* Fibrous calcite.

2. **Orthorhombic carbonates.** The common members are

(*i*) *Aragonite.* $CaCO_3$.   (*ii*) *Witherite.* $BaCO_3$.

(*iii*) *Cerrusite.* $PbCO_3$.   (*iv*) *Strontianite.* $SrCO_3$.

Physical properties :

(*i*) *Colour.* White, yellowish and grey.

(*ii*) *Lusture.* Vitreous but cerrusite shows admantine lusture.

(*iii*) *Hardness.* 5 in case of aragonite, 3 to 4 in case of cerrusite, witherite, and strontianite.

(*iv*) *Sp. gravity.* High in cerrusite (6), in others 3 to 4.

(*v*) *Cleavage.* Distinct in some cases and poor in others.

Optical properties. Nothing important.

Occurrence. Carbonates are usually secondary minerals, commonly occur as gangue minerals in ores of hydrothermal origin. But in sedimentary rocks they are quite important.

Important characteristics. All the carbonates react actively with acids.

PART V
PETROLOGY

CHAPTER 45

PETROLOGY
(*Introductory Concept*)

It is a branch of Geology, which deals with the study of rocks, and includes :

(*a*) *Petrogenesis, i e.,* origin and mode of occurrence as well as natural history of rocks.

(*b*) *Petrography, i.e.,* dealing with classification and description of rocks.

The branch of petrology dealing with the study of stones alone is called 'lithology'. Stones include the rocks that are necessarily hard, tough and compact.

As we know, rocks are necessarily the constituents of the earth's crust. Rocks are composed of minerals. Some rocks are monomineralic, composed of one mineral only while most of the rocks are multiminerallic consisting of more than one mineral species as essential constituents.

Igneous and metaigneous rocks constitute 95% of all the rocks of the earth's crust. Sedimentary and metasedimentary rocks constitutes 5% of the rocks of the earth's crust.

CLASSIFICATION OF ROCKS

According to the mode of origin, all rocks are categorised into three major groups :

I. Igneous Rocks or Primary Rocks.
II. Sedimentary Rocks or Secondary Rocks.

III. Metamorphic Rocks.

I. **Igneous rocks.** These are the rocks formed by the solidification of magma either underneath the surface or above it ; accordingly they are divided into two groups :

(*a*) *Intrusive bodies.* Which are formed underneath the surface of the earth.

(b) *Extrusive bodies.* These are due to the consolidation of magma above the surface of the earth. These are also known as Volcanic-rocks.

On the basis of the depth of formation, intrusive rocks are of two types :

(i) *Plutonic rocks,* which are formed at very great depths.

(ii) *Hypabyssal rocks,* which are formed at shallow depth.

Important Features of Igneous Rocks :

1. Generally hard, massive, compact with interlocking grains.
2. Entire absence of fossils.
3. Absence of bedding planes.
4. Enclosing rocks are baked.
5. Usually contain much feldspar.

II. Sedimentary rocks. These rocks have been derived from the pre-existing rocks, through the processes of erosion, transportation and deposition by various natural agencies like, wind, water, glacier, etc. The loose sediments, which are deposited, undergo the processes of compaction and the resulting products are known as sedimentary rocks.

On the basis of place of formation, sedimentary rocks are of two types :

(i) *Sedentary rocks,* that are the residual deposits, formed at the site of the pre-existing rocks from which they have been derived. These are not formed by the process of transportation.

(ii) *Transported,* in which case the disintegrated and decomposed rock materials are transported from the place of their origin and get deposited at a suitable site. According to the mode of transportation of the deposits, these rocks are sub-divided into three **types as :**

(a) *Mechanically deposited.*   Clastic rocks.
(b) *Chemical precipitation.*   Chemical deposits.
(c) *Organically deposited.*   Organic deposits.

Important Features of Sedimentary Rocks :

1. Generally soft, stratified, i.e., characteristically bedded.
2. Fossils common.
3. Stratification, lamination, cross-bedding, ripple marks mud-cracks, etc. are the usual structures.
4. No effect on the enclosing or the top and bottom rocks.
5. Quartz, clay minerals, calcite, dolomite, hematite are the common minerals.

III. Metamorphic rocks. These are formed by the alteration of pre-existing rocks by the action of temperature, pressure aided by sub-terranean fluids (magmatic or non-magmatic).

Important Features of Metamorphic Rocks :

1. Generally hard, interlocking grains and bedded (if derived from stratified rocks).

2. Fossils are rarely preserved in rocks of sedimentary origin except slates.

3. Foliated, gneissose, schistose, granulose, slaty, etc., are the common structures.

4. Common minerals are andalusite, sillimanite, kyanite, cordierite, wollastonite, garnet, graphite, etc.

A Nutshell of Classification

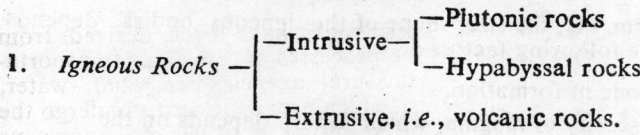

1. *Igneous Rocks* —

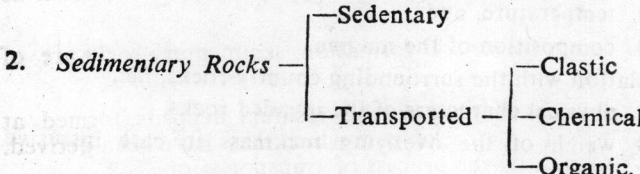

2. *Sedimentary Rocks* —

3. Metamorphic Rocks.

IMPORTANT ROCK TYPES OF INDIA

1. Igneous Rocks Types :

(*i*) Granite—with its volcanic equivalent, *i.e.*, Rhyolite.

(*ii*) Syenite—with its volcanic equivalent 'Trachyte'.

(*iii*) Nepheline-Syenite (and phonolite).

(*iv*) Anorthosite.

(*v*) Granodiorite and Monzonite.

(*vi*) Gabbro, diorite and norite, and their volcanic equivalents basalts (deccantraps), andesite etc.

(*vii*) Peridotite.

(*viii*) Carbonatite.

2. Sedimentary Rocks :

(*i*) Sandstone   (*ii*) Shale

    (*iii*) Limestone and dolomite   (*iv*) Saline rocks
    (*v*) Laterite.

3. Metamorphic Rocks :

(*i*) Gneiss	(*ii*) Schists
(*iii*) Quartzite	(*iv*) Khondalite
(*v*) Charnockite	(*vi*) Marble
(*vii*) Gondite	(*viii*) Kodurite
(*ix*) Slate	(*x*) Phyllites.

MODE OF OCCURRENCE
OR
FORMS OF IGNEOUS-ROCKS

The form, *i.e.*, the size, shape of the igneous bodies, depends mostly on the following factors :

    (*i*) Mode of formation.
    (*ii*) Viscosity of magma, which in turn depends on the
        (*a*) temperature, and
        (*b*) composition of the magma.
    (*iii*) Relation with the surrounding country-rocks, *i.e.*,
        (*a*) physical characters of the invaded rocks,
        (*b*) weight of the overlying rockmass in case intrusive bodies
        (*c*) structure.

The intrusive and the extrusive rocks exhibit typical forms, which are characteristic to them.

The forms assumed by intrusive bodies depend upon major geological structures as faults, folds, bedding planes, etc. Accordingly there are two major categories of forms of the intrusive bodies :

    1. Discordant-bodies.   2. Concordant-bodies.

1. Discordant-bodies. In this case an intrusive mass happens to cut across the structures of the pre-existing rocks of the country. There are different types of discordant forms in unfolded regions as well as in highly folded regions.

    (*a*) **In unfolded regions :**

    (*i*) *Dykes.* These discordant igneous bodies exhibit a cross-cutting relationship with the country rocks. Dykes commonly occur in groups and such group may be of radiating, arcuate or any other pattern.

Since for the formation of dykes the magma is to be sufficiently mobile, the composition of the dykes are mostly basic, *i.e.*, *doleritic.* Dykes are evidences of regional tension in the crust within the area of igneous activities. Larger dykes produce baking and hardening effect on either side.

(*ii*) *Ring dyke.* A dyke of arcuate out crop ; occurring more or less in the form of a complete or nearly complete circle.

(*iii*) *Cone-sheets.* These are inwardly dipping (in the form of inverted co-axial cones) dyke-like masses with circular out crop.

(*b*) **In high folded regions :**

(*i*) *Batholiths.* These are the largest intrusive bodies. Most batholiths are found in belts of deformation within the earth's crust and are granitic in composition. These are widening downwards to unknown depths.

Batholiths of comparatively smaller dimensions are called '*stocks*' and stocks of circular outcrop upon the surface are known as *bosses*. The remnants of the country-rock occurring upon or near the top surface of such intrusive masses are known as '*roof-pendants*'.

(*ii*) *Ethmolith.* These are funnel-shaped basic bodies with circular outcrop.

(*iii*) *Harpolith.* Sickle-shaped basic bodies formed by stretching of the strata after or during injection.

(*iv*) *Chonolith.* Any irregular intrusive body.

Any deep-seated intrusive body, irrespective of its shape and size, is known as '*pluton*'.

2. Concordant bodies. These are intrusive bodies that run parallel to the structures of the country-rocks in which they occur.

(*a*) **In unfolded regions :**

(*i*) *Sills.* These are thin parallel sided tabular sheet of magma that has penetrated along bedding planes, planes of schistosity, unconformities, etc. These are also doleritic in composition. They may attain any orientation in space depending upon the attitude of the rock beds in which they occur.

(*ii*) *Laccolith.* These intrusive bodies have their lower surface flat and have a convex top. It is due to accumulation of viscous magma which is usually acidic in composition, which pushes the overlying rocks upwards to make room for the mass.

(*iii*) *Lopoliths.* These are saucer shaped bodies, which are of huge dimensions and are of basic to ultrabasic in composition.

(*iv*) *Bysmalith*. Sometimes the magma breaks through the overlying rock beds and the igneous mass after consolidation is known as bysmalith.

(*b*) **In highly folded regions :**

(*i*) *Phacoliths*. These are crescentic shaped igneous bodies occurring along the crests and troughs of folds of the country rocks. They are basaltic in composition.

Forms of Extrusive Bodies :

(*i*) **Concordant**. Lava flows and pyroclastics which are the products of volcanic activities are the usual forms of extrusive igneous bodies.

(*ii*) **Discordant :**

Valcanic neck. It is a mass of igneous rock produced by the consolidation of lava and pyroclastic materials in the channel of eruption of an extinct volcano.

STRUCTURE OF IGNEOUS ROCKS

The structures of igneous rocks are large scale features, which are dependent on several factors like :

(*a*) Composition of magma.

(*b*) Viscosity of magma.

(*c*) Temperature and pressure at which cooling and consolidation takes place.

(*d*) Presence of gases and other volatiles.

Igneous structures are mostly classified into three major groups, as follows :

1.   **Mega-structures.**   2.   **Minor structures.**
3.   **Micro-structures.**

1.   Mega-structures. These are usually formed in the flow stage of the magma (*i.e.*, in the extrusive rocks), and incluae :

(*i*) **Vesicular and amygdaloidal structures.** When lavas heavily charged with gases and other volatiles are erupted on the surface, the gaseous constituents escapes from the magma as there is a decrease in the pressure. Thus, near the top of flows, empty cavities of variable dimensions are formed. The individual openings are known as vesicles and the structure as a whole is known as *vesicular structure.*

If, however, the vesicles thus formed are subsequently filled in with some low-temperature secondary minerals, such as calcite' zeolite, chalcedony etc., these infillings are called '*amygdales*'

Lavas containing amygdales are said to have amygdaloidal structure. These are the usual characteristic of basic lava flows.

(*ii*) **Cellular or scoriaceous structure.** By the bubbling out of the gases, from a lava heavily charged with volatile and gaseous constituents, numerous cavities are formed with the solidification of the lava. When the cavities are very much abundant, the term '*pumice*' or 'rock-froth' is applied. Such structures are known as cellular or scoriaceous structures and are characteristic of highly siliceous lavas.

(*iii*) **Lava-drain tunnels.** Sometimes while the upper surface of the lava consolidates, the interior may still remain fluid. When the enclosed fluid lava drains out through some weak-spots lying at the periphery of the flow, the resulting structure is known as lava-drain tunnel.

(*iv*) **Block-lava.** Since lavas of acidic composition, due to their high viscosity, do not flow to greater distances, they after solidification are found to offer a very rough surface. Such lava flows are known as block lava. It is also known as 'aa' structure.

(*v*) **Ropy lava.** Lavas of basic composition are quite mobile because of their low viscosity and they can flow to greater distances and after solidification offers very smooth surface. Such lava flows are known as ropy lava and are also known as 'pahoehoe' structure.

(*vi*) **Pillow structure.** It consists of isolated pillow shaped masses piled one upon another.. These are produced by extrusion of lava into rain-soaked air, beneath ice-sheets, under water logged sediments or in sea water. Spilite, a lava rich in albite (*i.e.*, sodium rich) characteristically exhibits pillow structure.

(*vii*) **Sheet structure.** The development of one set of well defined joints, sometimes brings about a slicing effect on the massive igneous rock body. If all such slices are horizontal, the structure is said to be sheet structure.

(*viii*) **Platy structure.** This is also due to the development of different sets of joints, which gives rise to only plates of the rock mass, on striking the rock. Such a feature is known as platy structure.

(*ix*) **Columnar structure.** As a consequence of contraction due to cooling, a few sets of vertical joints develop. Such joints, bring about the formation of columns, which may be square, rectangular, rhombic or hexagonal in outline.

(*x*) **Flow structure.** Subsequent to eruption of lava upon the surface the viscous, varieties flow from one place to the other with great difficulty and in their attempt to do so, the dissimilar patches within the lava are drawn out in the form of elongated lenticles. Sometimes the already crystallised particles within the magma are arranged parallel to the direction of flow of the lava. They naturally indicate the direction of flowing of the mass, prior to its consolidation. These are also known as directional structure or more commonly flow structure.

(*xi*) **Rift and grain.** These are due to jointing. In granites, three mutually perpendicular, equally spaced joints, which are taken into advantage while producing cubical blocks, are known as 'mural jointing'. But for processing of the blocks down to smaller dimensions, the mutually perpendicular closely spaced joints (one horizontal and the other vertical) are taken into advantages. These joints are known as rift and grains.

2. Minor structure. These structures are formed in the fluid stage of the magma (*i.e.*, in the intrusive rocks) and include the following :

(*i*) **Primary foliation.** Sometimes many plutonic rocks are characterised by foliation resulting from the parallel arrangement of platy and ellipsoidal mineral grains.

(*ii*) **Banding in rocks.** These are also known as layered rocks consisting of alternating bands of different composition. It may result from lamellar flow, from settling of minerals from a crystallized magma or from successive injections.

(*iii*) **Schlieren.** These are somewhat wavy, streaky, irregular sheets, usually lacking sharp contact with the surrounding igneous rocks. They may be altered inclusions, segregation or may represent concentration of residual fluids into layers in a rock that had otherwise crystallised.

3. Micro structures. These are formed due to reaction between already solidified crystals and the rest of the magma and include the following :

(*i*) **Reaction rims.** When the reaction between an already crystallized mineral and the rest of the magma is incomplete, the corroded crystals are found surrounded by the products of reaction *i.e.*, some new mineral. Such zones are known as reaction-rim. When the reaction rims are produced by primary magmatic reaction, they are known as '*corona structures*' and '*kelyphitic borders*' when secondary.

(*ii*) **Myrmekite structure.** It is produced by an intergrowth of quartz and plagioclase feldspar where quartz occurs as blebs or drops in plagioclase.

(*iii*) **Graphic structure.** It results from an intergrowth of quartz and orthoclase feldspar.

(*iv*) **Xenolithic structure.** Occurrence of foreign rock fragments within an igneous rock gives rise to xenolithic structure. The xenoliths are said to be *'cognate'* when they are genetically related to enclosing rocks and *'accidental'*, when they are fragments of country-rocks without having any genetic relation with the enclosing rock.

(*v*) **Orbicular structure.** These are spherical segregations consisting of concentric shells of different mineral composition and texture, which occasionally occurs in granitic rocks.

(*vi*) **Spherulitic structure.** Its essential feature is simultaneous crystallization of fibres with radiating arrangement about a common centre. The large spherulites are known as 'Lithophyse'. In basic lavas and intrusions, they are called *'varioles'* and the rocks containing them variolites.

(*vii*) **Perlitic-cracks.** These are curved, concentric lines of fracture, often seen in volcanic glass. These are simply due to contraction of the glassy mass on cooling.

TEXTURE OF IGNEOUS ROCKS

Textures of igneous rocks describe the actual relations between crystals or that between the crystals and the glassy material present within igneous rocks. Rocks have been formed under diverse physicochemical environment, and textural studies indicate the cooling history of the magma. Texture of igneous rocks is a function of three important factors :

(1) degree of crystallisation, *i.e.*, crystallinity.
(2) size of the grains (or crystals), *i.e.*, granularity.
(3) fabric, which includes :
   (*i*) shape of crystals.
   (*ii*) mutual relations of grains or of crystals and glassy matter.

1. Crystallinity :

(*a*) *Holocrystalline.* When an igneous rock is made up of mineral grains only.

(*b*) *Hemi-crystalline.* When a rock contains both crystalline as well as glassy matter in variable proportions.

(*c*) *Holohyaline.* When the igneous rock consists wholly of glass.

The degree of crystallisation depends on the following factors:

(*i*) Rate of cooling.
(*ii*) Viscosity of magma (*i.e.*, composition of the magma and the presence of volatile components).
(*iii*) Depth of cooling.
(*iv*) Volume of the magma.

2. Granularity. It refers to the grain size of the crystals present in the igneous rocks. **These are**

(*a*) **Phaneric.** When individual crystals are visible to the naked eyes and are

(*i*) *Coarse grain.* When the grain size is 5 mm or above.

(*ii*) *Medium grain.* Grain size is 1 mm to 5 mm.

(*iii*) *Fine grain.* Grains are smaller than 1 mm in diameter.

(*b*) **Aphanitic.** When individual grains cannot be distinguished with unaided vision and are

(*i*) *Micro-crystalline.* When individual crystals are distinguishable only under microscope.

(*ii*) *Mero-crystalline.* Intermediate in range.

(*iii*) *Crypo-crystalline.* When individual crystals are too small to be separately distinguished, even under the microscope.

(*iv*) *Glassy.* When there is no crystallisation at all.

In general, pegmatitic and plutonic rocks are coarse grained, hypabyssal rocks are medium grained and volcanic rocks are either fine grained or glassy.

In most natural glasses there are found number of minute bodies of various shapes, which represent the beginnings of crystals. They are

(*a*) *Crystallites.* These are embryo crystals.

(*b*) *Microlites.* Minute crystals, which may exhibit the crystal outline appropriate to their mineralogical nature.

Terms like globulites, margarites, longulites, trichites, scopulites are used to describe various forms of microlites.

Devitrification. It is the process of conversion of glassy material to crystallised state. Perlitic cracks in crystals are evidences of their original glassy condition.

3. Fabric :

(*i*) **Shape of the grains.** It refers to the degree of development of crystal faces and are :

(*a*) *Euhedral.* When the mineral grains are found to have developed a perfect crystal outline. These are known as *idiomorphic* or automorphic crystals.

(*b*) *Sub-hedral.* When the crystal outlines have partially developed. These are also known as hypidiomorphic or hypautomorphic crystals.

(*c*) *Anhedral.* When the crystal faces are absent.

With reference to the three dimensions in space, crystals are classified as

(a) *Equidimensional.* Crystals found to have developed equally along all directions in space.

(b) *Tabular.* When better developed in two directions.

(c) *Prismatic.* When better developed in one direction only.

(ii) **Mutual relationship.** It refers to the relative size, shape and dimensions of crystals and their relation to one another. This may be :

(a) Equigranular, and   (b) Inequigranular.

(a) **Equigranular :**

(i) **Allotriomorphic.** In this case all the crystals are anhedral. Synonymous terms are xenomorphic, aplitic, mosaic, sugary, sacchraoidal.

(ii) **Hypidiomorphic.** Here all the crystals are subhedral.

(iii) **Panidiomorphic.** Here all the crystals are euhedral. This is the characteristic texture of lamprophyres.

In case of aphanitic rocks, they are

(a) *Micro-granitic.* Consisting of anhedral and subhedral grains.

(b) *Orthopyric.* Crystals are microscopically euhedral.

(c) *Felsitic.* Grains are too small to be identified in thin sections.

(b) **Inequigranular.** In this case the grain size shows a marked difference from grain to grain. This is also known as seriate texture, and are of the following types :

I. (i) **Porphyritic texture.** In this case larger crystals are enveloped in a groundmass which may be micro-granular, merocrystalline or even glassy. This texture is characteristic of volcanic and hypabyssal rocks. Its origin may be attributed to :

(a) Change in physico-chemical condition.

(b) Molecular concentration.

(c) Insolubility.

(ii) **Glomero-porphyritic.** Here phenocrysts gather at one spot.

(iii) **Vitrophyric.** When the groundmass is glassy in a porphyritic texture, it is called vitrophyric texture.

(iv) **Felsophyric.** Here groundmass is crypto crystalline in nature.

II. (i) **Poikilitic texture.** In this case, smaller crystals are enclosed in the larger ones without common orientation. The enclosing crystal is known as '*oikocrysts*' and the enclosed ones are called 'chadacrysts.'

(*ii*) **Ophitic texture.** It is a type of poikilitic texture in which euhedral plagioclase crystals are partially or completely surrounded by anhedral augite crystals. This is the characteristic texture of dolerites.

(*iii*) **Sub-ophitic texture.** With equal size of feldspar grains and augite, the enclosure is partial and it is known as sub-ophitic.

(*iv*) **Granulo-ophitic.** Here grains of plagioclase are enclosed within a large patch of pyroxene, and the latter in turn is made up of a number of pyroxene grains.

(*v*) **Hyalo-ophitic.** When diversely oriented grains of plagioclase occur within the glassy groundmass.

(*vi*) **Ophimottling.** Ophitic texture confined to separate areas.

III. Intersertal and intergranular texture. When the plagioclase laths are arranged in a tringular fashion and the polygonal interspace left between the crystals is having glassy infillings, it is known as '*intersertal*'. However, if the interspace is filled in with mineral grains, it is called '*intergranular*'. These are commonly found in basalts.

IV. Directive textures. These are produced by flow in magma during their crystallisation.

(*i*) **Trachytic texture.** Produced by subparallel arrangement of plagioclase along the direction of flow of lava. Without evident flow, it is known as Felsitic texture.

(*ii*) **Hyalopilitic.** Also known as felty, which is due to appreciable admixture of glass and feldspar needles within the body of a rock.

(*iii*) **Intergrowth :**

(*a*) Graphic intergrowth between orthoclase and quartz.

(*b*) Perthitic intergrowth between albite and orthoclase.

(*c*) *Myrmekite growth* between quartz and feldspar.

(*d*) *Granophyric* is the graphic texture found in hypabyssal and volcanic rocks of granitic composition.

As described above, these are important and common textures of igneous rocks.

CHAPTER 46

MAGMA

(*Its Composition and Constitution*)

Magma is a natural rock fluid beneath the earth's crust, which may consolidate to form and igneous rock. When magma is erupted to the surface, it is known as lava, the consolidation of which gives rise to volcanic rocks.

Composition and Constitution. Magmas consist of mixtures of solids, fluids and dissolved gases. Essentially they are very hot silicate melts containing large quantities of water and varying amounts of highly reactive fluids and gases in solution. These reactive fluids include such things as hydrochloric acid and hydrofluoric acid. Magmas do not have a fixed composition. Although the compositions of different magmas undoubtedly vary, many are close to the following composition :

(*a*) **Chemical composition :**

(*i*) In terms of oxide :

SiO_2	59%	Al_2O_3	15%
Fe_2O_3	3%	FeO	3·5%
CaO	5%	Na_2O	3·8%
MgO	3·5%	K_2O	3%
H_2O	1%	TiO_2	1%
P_2O_5	0 3%	MnO	0·1%
CO_2	0·1%.		

(*ii*) **In terms of elements.** Oxygen, Silicon, Aluminium, Iron, Calcium, Sodium, Potassium, Magnesium, Titanium, etc. They together constitute more than 99% of the fixed constituent of any magma.

(b) **Mineralogical composition :**

Feldspars	59%
Pyroxene and Amphibole	17%
Quartz	12%
Mica	4%
Others	8%

Through the processes of differentiation and assimilation, the same magma may give rise to diverse rock types.

Crystallization of magma. It is well known that magmas become igneous rocks by solidification without crystallization (*i.e.*, formation of glass) or by crystallization with the loss of much of their volatile materials. The crystallization of magma is governed by several factors like :

(i) Temperature (rate of cooling).

(ii) Viscosity of magma.

(iii) Composition of magma.

(iv) Concentration of volatiles.

(v) Pressure (depth of cooling).

Magmas usually consist of a number of components and therefore most igneous rocks are multi-component. Unicomponent rocks are extremely rare. The crystallization can be better studied in a unicomponent, bi-component and multi-component system.

Unicomponent-magma. In the study of unicomponent magma, the temperature region in which the generation of crystals is slow is called the metastable region and the region in which the rate of crystallization is rapid is the labile region.

According to this study, it has been observed that slow-cooling leads to coarse grains whereas rapid-coolings leads to the formation of glass. Besides, multi-component rocks are of finer grains than the one of simpler composition.

Bi-component magma. The important principle, *i.e.*, "melting point of any of the components in a bi-component system is lowered-down to variable extents due to the presence of variable amounts of the other one" is the guiding principle in the crystallization of bi-component magmas. Bi-component magmas show two types of relationship and method of crystallization as

1. Eutectic crystallization.

2. Solid-solution or Mixed-crystals.

Eutectic relation. In this case the two components having distinct and different freezing points, shows decrease of their freezing points at various stages of their combination. At a particular point of their combination, and at a particular temperature these two components crystallize together. This temperature is called the 'Eutectic temperature' and the point of combination showing a specific percentage of both the components in the mixture, (*i.e.*, composition) is called the Eutectic-point.

This type of crystallization gives rise to the intergrowth of the two mineral and results in a peculiar graphic-texture.

    e.g.,   *Perthite.* Orthoclase : Albite 42 : 58

            Graphic. Orthoclase : Quartz 72·5 : 27·5.

Solid solution. Sometimes, it is observed that both the components of a binary magma are isomorphous and miscible in all proportions in the solid-state forming homogeneous crystals. The best examples are plagioclases and pyroxenes.

In a temperature composition diagram (to illustrate this type of crystallization) there are two curves—solidus and liquidus. The melting point curve is called the solidus and the freezing point curve is called the liquidus.

Such crystals do not melt at a definite temperature but melting is spreading over a range of temperature, the lower limit is fixed by the solidus and the upper limit by the liquidus.

If a mixture of the two-components having a particular percentage of their combination start cooling down, the crystallization will only start when a particular temperature is reached.

If a magmatic mixture of composition '*P*' is cooled to 1400°C, the crystals of composition '*Q*' will begin to form. As the temperature continues to fall the liquid gets enriched with the albite (Ab) component . by the withdrawl of anorthite (An) rich crystals.

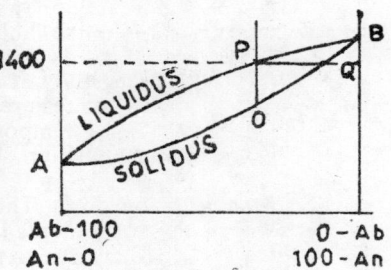

Fig. 46·1.

In this way, the continuous crystallization of solids, their reaction with the melt and consequent changing over to newer solids of ever changing composition continue till the composition of the solid phase is the same as that of the original composition of the melt.

In the case of such solid-solution relationship, with rapid cooling the crystals do not change to a different composition, in consonance with the rate of cooling and therefore zones of different composition appropriate to the temperature (at which it was formed) are formed around the early formed crystals. The examples are best found in plagioclase crystals.

This is also known as continuous reaction-relation.

Besides the above, binary magmas also sometimes show peritectic relationship, in systems like forsterite-silica, etc.

Tri-component magma. These are ternary magmas in which the relationship between all the three components may include both solid-solution and/or eutectic characteristics.

Thus the crystallization of magma has been explained by various geologists through laboratory experiments.

CHAPTER 47

BOWEN'S REACTION PRINCIPLE

According to the studies made by Dr. N.L. Bowen on the crystallisation of and the reactions which take place in a basaltic magma as it is cooled, he enunciated the reaction principle. Bowen's reaction principle illustrates how a magma may solidify as a single rock-type or may give rise to many rock types.

As crystallisation of magma proceeds there is a tendency for equilibrium to be maintained between the solid and liquid phases To maintain this equilibrium, early formed crystals react with the liquid and changes in composition takes place. The sequence of crystallisation as has been worked out by Bowen (1922), is known as 'Bowen's reaction series'. In this series, the minerals are so arranged that each is supposed to react with the magmatic fluid so as to produce the one placed below it.

In fact, there are two parallel series—one series to represent the transformation in structure and composition of the ferro-magnesian minerals with the falling temperatures and the other series represents the cooling and crystallisation of minerals of plagioclase group in the magmatic fluid with the decreasing temperature. The reaction for the ferromagnesian minerals is called the 'Discontinuous series' and that of the plagioclase minerals is called the 'Continuous Reaction Series'. Both these two series converge and merge into a single series, which also forms a part of the discontinuous reaction series. The minerals in the reaction series indicate, in general, the order in which each mineral crystallises from a cooling basaltic magma.

Unless the early formed crystals are removed from the melt or if its composition cannot be further changed (*i.e.*, magmatically dead), reaction between the magma and the crystals formed will take place to produce new minerals.

The temperature range, for which the Bowen's Reaction Series has been worked out, is from 1100°C to 573°C.

SPINELS

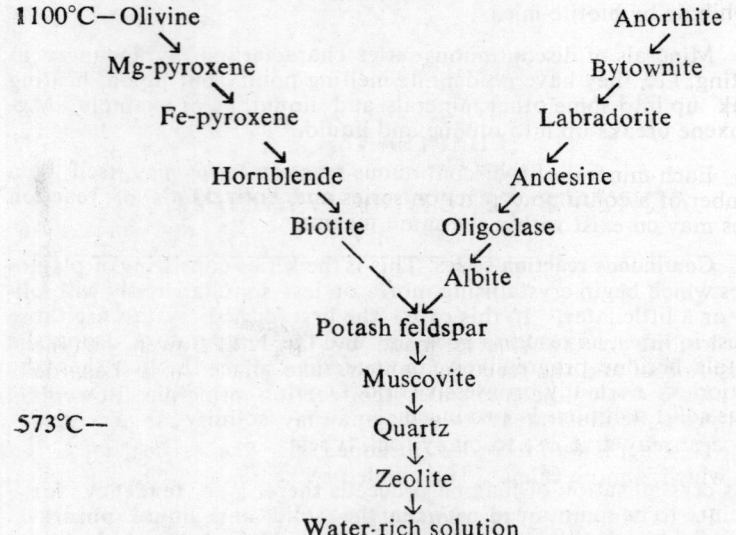

After the crystallization of the minerals of the Spinel-group, at a temperature of about 1100°C, the first signs of crystallization of silicate minerals begin to take place.

Discontinuous reaction series. The first mineral to crystallize in the Discontinuous Reaction-series is Mg-olivine. As soon as olivine is formed and unless removed from the seat of action, reacts with the magma so as to produce a mineral with which the phase is saturated under the existing temperature. Thus with the falling temperature olivine is converted into Fe-divine and then to Mg-pyroxene (clino-enstatite). This will continue till the whole of olivine is converted into pyroxene, under ideal physico-chemical condition. The two minerals thus related by reaction are called a *'reaction-pair'*.

In this way, Mg-pyroxene gets converted into Ca-pyroxene and then to amphibole (hornblende) which in turn gets transformed to biotite, with the falling temperature.

The greater the degree of fractionation, the more extensive is the reaction process. When the reaction between crystals and liquid goes into completion, the minerals of the final rock are obviously those formed late in the series ; as the early formed minerals were dissolved and absorbed during the reaction. But if the reaction is incomplete because of too rapid cooling or other reasons, early members of both reaction series may remain as relics in the final rocks and reaction rims are formed surrounding the early formed crystals. Thus olivine is surrounded by pyroxenes, pyroxene by amphibole and amphibole by biotite-mica.

Minerals of discontinuous series characterised by incongruent melting, *i.e.*, they have no definite melting point, but upon heating break up into some other minerals and liquid. For example, Mg-pyroxene breaks up into olivine and liquid.

Each mineral of a discontinuous reaction series may itself be a member of a continuous reaction series and both kinds of reaction series may co-exist within the same magma.

Continuous reaction series. This is the series consisting of plagioclases which begin crystallising more or less simultaneously with olivine or a little later. In this case, the first formed crystals are those richest in lime ; as reaction goes on and the temperature drops, the crystals became progressively more sodic. This implies that the reaction is normally progressive and a continuous series of homogeneous solid solutions is produced.

This fact is well recorded in zoned plagioclases in which the core which is more calcic is surrounded by successive soda rich zones.

Partial failure of reaction between olivine and liquid results in the enrichment of the liquid in silica, and the final crystallised product may be a mixture of olivine, pyroxene and quartz. In this case the quartz is called a *released mineral*. Rocks containing released minerals are called *doliomorphic*.

Importance of Bowen's reaction principle :

(1) It illustrates how a magma may solidify as a single rock type or may give rise to many rock types. The primary basaltic magma may solidify as a gabbro consisting of olivine and calcic plagioclase or it may give rise to rocks varying from dunite through gabbro, diorite, tonalite, granodiorite to granite, depending upon the degree of fractionation and the extent to which the early formed minerals are removed from further reaction with the melt.

(2) The atomic structure becomes more complicated from early formed minerals like olivine to the minerals like quartz, zeolite etc.

(3) The early formed crystals are more dense than the late-formed minerals of the Bowen's reaction series.

(4) It indicates the process of fractional differentiation in magma.

CHAPTER 48

DIFFERENTIATION

It is widely accepted that there is only one parental magma of basaltic composition and all the different varieties of igneous rocks were supposed to have originated from this magma of uniform composition. The origin of diverse igneous rocks with regards to mineralogical composition and texture can be attributed to two causes :

    I.   Differentiation   II. Assimilation

I. Differentiation. It may be defined as "the process whereby, a magma originally homogeneous splits up into contrasted parts, which may form separate bodies of rocks or may remain within the boundaries of single unitary mass". The process of differentiation, is usually favoured by two factors :

    (*a*) Rate of cooling.

    (*b*) Settling of early crystallized heavy minerals.

Stages of differentiation. According to Tyrrel there are two stages, in the first stage, there is preparation of units such as crystals, liquid submagma etc. In the second stage the prepared units are separated and accumulate separately to form distinct masses.

Differentiation in an igneous magma involves processes like :

    1. Fractional crystallisation.   2. Gravity separation.
    3. Filter pressing.   4. Liquid immiscibility.
    5. Gaseous transfer.

1. Fractional crystallisation. With the cooling of the magma, crystallisation begins and earliest minerals start crystallising. Differentiation may be brought about by at least two distinct processes :

    (*a*) The localisation of crystallisation aided by diffusion and convection.

(*b*) The localised accumulation of crystals in several different ways, with the concomitant segregation of the liquid magmatic residuum.

Crystallisation may be localised at a cooling margin, where the temperature is lower than the central parts of the magma. Thus two phases—a solid and a liquid are formed.

The concentration of the molecules of the growing crystals at the site of crystallisation is supposed to be due to (*a*) free ionic diffusion of that substance from all parts of the magma, (*b*) by convection current with a concomitant movement of other substances in the opposite direction. But these suppositions were later on found untenable.

During crystallisation, there is a tendency for equilibrium to be maintained between the solid and liquid phases. To maintain equilibrium, early formed crystals react with the liquid and changes in composition take place. In case of plagioclase, for instance, the first formed crystals are those richest in lime ; as reaction proceeds with falling temperature, the crystals become progressively sodic. Thus a continuous series of homogeneous solid solution is produced, which constitute the 'continuous-reaction series'.

Certain ferromagnesian minerals on the other hand react with the melt to give rise to a new mineral with a new crystal structure and a definite composition. Olivine, for example, may be transferred to pyroxene, and pyroxene to amphibole. Such abrupt changes constitute the discontinuous reaction series.

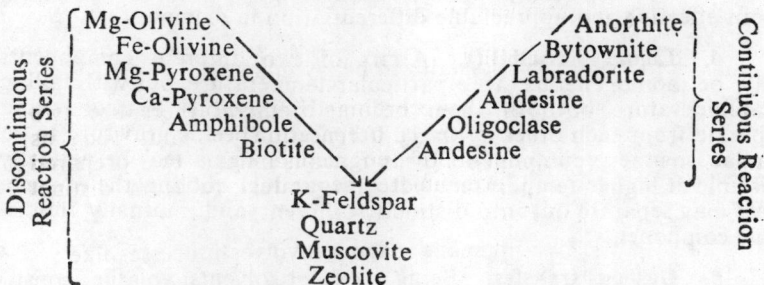

Certain minerals in igneous rocks are associated because they crystallize over the same range of temperature. Early high-temperature minerals of both series generally crystallize together. As a result while some minerals are characteristically associated with some specific minerals, others are incompatible with them.

'Bowen's Reaction Principle' illustrates how a primary basaltic magma may solidify as a gabbro or it may give rise to rocks varying from dunite through gabbro, diorite, tonalite, granodiorite to granite depending upon the degree of fractionation and the extent to which early formed minerals are removed from further reaction with the melt.

Thus two magmas of identical initial composition but cooling at different rate produces different rock types. In the absence of volatiles the normal minerals of the discontinuous reaction series cannot form.

The product of early crystallisation are concentrated at one end of a differentiation series and the products of later crystallisation, at the other end.

2. Gravitational settling. It is the tendency of the heavy minerals to sink to the bottom and those having lower specific gravity than the melt rise up and float at the top of the magma chamber. The perfection of this process depends on the size, shape and specific gravity of individual crystals and also on the viscosity of the magma. Olivine, seems to be the most important mineral affected by this process and its gravitational settling forms stratification in igneous rocks.

3. Filter pressing. As crystallisation continues a loose mesh or frame-work of crystals with residual liquid in the interstices will ultimately be formed. If, at this stage, deformation of the mass occurs, either by the lateral earth pressure or downward pressure of the lifted strata, the interstitial liquid will be squeezed out. The liquid will tend to move towards the region of least pressure. Thus, this process of separation of solid crystals from the fluid magma is known as *filter-pressing* and is found to be very helpful in bringing about effective and appreciable differentiation in magma.

4. Liquid immscibility. A mix of two different components may be homogeneous at a particular temperature, but with falling of temperature both of them become immiscible fractions and separate from each other by the difference in specific gravity. In a similar manner, components of an igneous magma may be perfectly miscible at higher temperature but with gradual cooling the magma mass may separate out into distinctly different and mutually immiscible components.

5. Gaseous transfer. Being excellent solvents, volatile constituents continually go on collecting the otherwise sparsely disseminated metallic and non-metallic constituents as they rise upwards through the magma chamber. Again the escaping gas bubbles may attach themselves to growing crystals and float them upwards. The volatile constituents are capable of making selective transfer of material from lower to higher levels. In this way, pronounced heterogenity may develop in magma.

Thus, differentiation is a major process that is responsible for bringing about diversity in igneous rock masses.

CHAPTER 49

ASSIMILATION

Assimilation is an important factor in bringing about diversity in igneous rocks. This is the process whereby rockmasses are incorporated by magmas, there is also commingling of two liquid magmas. Since these processes involve the re-mixing of rocks, they represent the reverse of the differentiation processes and heterogenity results when the mixing is incomplete and non-uniform.

The laws of assimilation are governed by the same general laws of fractional crystallisation. Reaction between magma and wall-rock is a normal accompaniment to igneous intrusion. In the course of this reaction the magma becomes contaminated by incorporating materials originally present in the wall-rock. This broad process of modification is described as assimilation.

The incorporation of foreign rock matter by a magma occurs in three ways as

(a) Mechanical incorporation without chemical reaction.

(b) Reactions involving partial solution of the incorporated matter and the precipitation involving the replacement of one solid phase by another.

(c) Total dissolution involving total disappearance of the solid phase.

In general it is a complex process of reciprocal reaction between magma and invaded rock. During the process of reaction, due to ionic exchange between liquid and crystals, minerals are changed into those crystalline phases with which the liquid was already saturated. The end product is a contaminated igneous rock which was at no time entirely liquid and which is made up of materials contributed partly by the original rock and partly by the wall-rock. The rocks formed in this way is naturally of hybrid origin, which are particularly common along the borders between intrusive and invaded rocks.

Factors affecting assimilation :

(i) Temperature of the magma at the time of intrusion.

(*ii*) Presence or absence of notable degree of superheat, *i.e.*, the stage at which the inclusions are tapped in.

(*iii*) Composition of the inclusions.

(*iv*) Concentration of volatiles in the magma.

(*v*) Conditions which facilitate or retard the escape of volatiles into the surrounding-rocks.

Since the melt reacts with the minerals which are formed earlier at a higher temperature, and gives rise to the minerals which at the moment are in equilibrium ; as in Bowen's Reaction Series. These reactions are exothermic, that is, they proceed with the production of heat and not the absorption of it. Only those inclusions made up of minerals belonging to the lower series can be directly dissolved ; the heat required for dissolving the inclusions is supplied by the crystallisation of a thermally equivalent quantity of those phases with which the magma at that moment is saturated.

According to the above observations, a general rule has been enunciated (by Zirkel) "in acid magma acid inclusions are not assimilated but basic ones are ; likewise in basic magma basic inclusions are not digested but acid ones are".

There are some petrochemical considerations underlie the reaction between magma and wall rock :

(*a*) Suppose, a magma of granitic composition has started reacting with the wall-rock of gabbroic composition ; in such a case labradorite and augite of gabbro are earlier members than oligoclase and hornblende of granitic magma. A complex reaction takes place whereby the minerals from the walls of gabbro are changed into hornblende and oligoclase, minerals which are in equilibrium with the melt at that particular temperature. This is the assimilation of basic igneous rocks by acid magma.

(*b*) **Assimilation of acid inclusions by basic-magma.** Basaltic magma is capable of melting acid igneous rocks, as its temperature is much above the melting point of acid igneous rocks. In such cases the members of the late-crystallization go into solution by the magma. To supply the necessary heat for fusion and reaction, an equivalent amount of the members of early crystallization get precipitated from the liquid.

(*c*) **Assimilation of sedimentary-rocks by basic magma.** Since sedimentary rocks are mostly composed of quartz, alkali-feldspar, clay minerals and calcite which are low-temperature minerals, they are completely incorporated by basic magmas.

Assimilation of calcareous inclusions desilicate the magma by crystallizing out various lime silicates as melillite, garnet etc. giving rise to a silica poor alkaline-residue and cause felspathoids to appear.

Incorporation of argillaceous matter may give rise to cordierite, sillimanite, spinel, garnet, anorthite etc.

Partial digestion of siliceous inclusions show reaction rims of augite, hypersthene, cordierite, plagioclase etc.

(*d*) **Assimilation of sedimentary-rocks by acid-magma.** It leads to a parallel development of minerals in the xenolith and in the magma.

(*e*) **Assimilation of metamorphic-rocks by basic magma.** Since metamorphic rocks are either altered igneous or sedimentary rocks, the effect of assimilation of such inclusions are guided by their temperature of formation as well as mineralogical composition.

CHAPTER 50

CLASSIFICATION OF IGNEOUS ROCKS

Igneous rocks show great variations both in chemical and mineralogical composition as well as textural characteristics. There is no general agreement among the petrologists as to the classification of igneous rocks. Different classifications have been proposed on the basis of

1. Mineralogical composition.
2. Chemical composition.
3. Textural characteristics.
4. Genesis.
5. Association.

1. On the basis of mineralogical composition. Since the relative amounts of the various minerals in a rock can be measured or estimated with a fair degree of accuracy, this criteria is given more emphasis for the classification of igneous rocks.

The minerals occurring in igneous rocks may be classed as

(*a*) Essential, (*b*) Accessory, and (*c*) Secondary.

The first two are the products of magmatic crystallisation whereas the third one is formed by the alteration of the primary minerals, *i.e.* (the 1st two), through the processes of weathering, metamorphism etc.

(*a*) **Essential minerals.** These are the major constituents of the rock which help in the diagnosis of the rocks types. The disappearance of these minerals would cause the relegation of the rock to another type.

(*b*) **Accessory minerals.** These are the minor constituents of a rock and their presence has no bearing on the nomenclature of the rock.

The minerals commonly occurring in igneous rocks may be broadly classified into *felsic* and *mafic* varieties. Felsic is a term derived from feldspar, felspathoid and silica and *mafic* is similarly derived from ferromagnesian minerals, like biotite, pyroxene, amphiboles etc.

Felsic minerals are light in colour, low melting point, low specific gravity, comparatively of late crystallisation.

The mafic minerals are dark coloured, higher specific gravity, higher melting point and are early crystallised minerals.

Terms like *leucocratic* and *melanocratic* refer to light coloured and dark coloured minerals respectively. Usually the felsic minerals are leucocratic and the rocks containing more of mafic minerals are melanocratic. The rocks which are pitch dark in colour are termed as '*hypermelanocratic*.' Mesocratic rocks are intermediate in colour.

2. Chemical composition. The most important chemical classification have been proposed by four American petrologists, Cross, Iddings, Pirsson and Washington. Here the chemical analyses are calculated into a set of hypothetical minerals called 'standard' or '*norm*', which are divided into two groups :

Salic Minerals	Femic Minerals
Quartz, orthoclase, albite, anorthite, leucite, nepheline corundum, zircon etc.	Acmite, diopside, apatite, wollastonite, hypersthene, olivine, sphene, ilmenite, magnetite, chromite etc.

On this basis the rocks are classified as follows :

1. Salic to femic minerals into five classes.

2. Quartz to feldspar and of accessory minerals to other femic minerals into orders.

3. Salic alkalies to femic alkalies to famic-limes into rangs.

In their chemical composition, the igneous rocks vary between wide limits. Thus rocks like granite may contain about 70 to 80% of silica and very little quantity of iron, magnesia and lime while at the other extreme end, there are rocks like peridotita, dunite, etc. which often contain about 35 to 40% of silica and larger quantities of iron, magnesia and lime. Thus igneous rocks may be classified as

    (*a*) Acid   (*b*) Intermediate

    (*c*) Basic   (*d*) Ultrabasic rocks.

(*a*) **Acid-igneous rocks.** These rocks have more than 65% of the silica content, *e.g.*, Granite, Grano-diorites.

(*b*) **Intermediate rocks.** These are having 55 to 65% of silica. *e.g.*, syenite and diorites.

(*c*) **Basic rocks.** Here the silica content is between 44 to 55%, *e g.*, basic.

(*d*) **Ultrabasic rocks.** In this case, silica content is less than 44%. Rocks like anorthosite belongs to this category. Ultrabasic rocks having higher magnesian contents are known as ultramafics.

Shands and Holmes classified the igneous rocks as

(*i*) **Supersaturated rocks.** Also known as oversaturated rocks. Here, the excess of silica crystallize as quartz.

(*ii*) **Saturated rocks.** They have just sufficient silica to form the stable silicate minerals but no free quartz.

(*iii*) **Under-saturated rocks.** They contain insufficient silica and minerals like olivine, nepheline, leucite etc. are therefore present.

3. Textural classification. It is mostly based on the cooling history of the magma and the modes of occurrence of these rocks. Accordingly there are (*a*) Plutonic (*b*) Hypabyssal, and (*c*) Volcanic rocks.

The plutonic rocks are formed under deep seated conditions, where the temperature and pressure are very high and the rate of cooling is very slow. Hence their texture is holocrystalline and coarse.

The hypabyssal group includes the rocks of dykes, sills and small laccoliths etc., which occupies intermediate position in the crust between the plutonic and volcanic rocks. Their texture is usually merocrystalline.

The volcanic rocks on the other hand are formed on the surface of the earth and due to rapid rate of cooling their texture becomes holohyaline and fine grained.

Recent classification. It takes into accout, factors like mineralogical composition, chemical and textural characteristics etc. and is represented in a tabular form. The important features are as follows :

(*i*) Presence or absence of essential feldspar—classes 1st and 2nd.

(*ii*) **Class—I** :

Group-A—Acid Igneous Rocks.

Group-B—Intermediate and Basic Rocks.

Group-C—Alkaline Igneous Rocks.

Group-D—Ultrabasic and Ultramafics.

(*iii*) In Group-'B', there are two sub-groups, one is characterised by essential alkali feldspar and the other by essential plagioclase.

In Group-'D', there are two ʃsub-groups on the basis of presence or absence of felspathoid.

(*iv*) **Class-II.** There are two sub-groups on the basis of presence or absence of felspathoids.

Thus the igneous rocks have been classified. A list of the various important igneous rocks and their plutonic, hypabyssal and volcanic equivalents, are as follows :

(*i*) *Granite*. Alkalifelspar predominant—Rhyolite.

(*ii*) *Adamellite*. Plagioclase : Alkalifelspar—Rhyodacite.

(*iii*) *Granodiorite*. Plagioclase predominant—Dacite.

(*iv*) *Syenite*. Alkalifelspar dominant—Trachyte.

(*v*) *Monzonite*. Alkalifelspar : Plagioclase—Trachyandesite.

(*vi*) *Diorite*. Plagioclase dominant—Andesite.

(*vii*) *Gabbro*. Plagioclase of labradorite type—Doterite (Hypabyssal)—Basalt (Volcanic).

(*viii*) *Nepheline syenite*. Felspar + Felspathoid—Phonolite.

(*ix*) *Anorthosite*. Only labradorite type of felspar—Limburgite.

(*x*) *Theralite*. Ultramafic with felspathoid—Tephrite.

(*xi*) *Ijolite*. Ultramafic with felspathod-felspathoid base.

CHAPTER 51

COMMON IGNEOUS ROCKE OF INDIA

1. **Granite.** In the Eastern-ghats, Rajasthan, in the Himalayan ranges, these rocks are in abundance. They mostly contain, ortho-clase, quartz, muscovite and small amounts of hornblende.

2. **Basaltic rocks.** As deccan traps, occur in abundance in Gujarat, Madhya Pradesh, Maharashtra, A.P. etc. Their mineralogic contents are usually plagioclase, augite sometimes a little olivine.

3. **Pegmatites.** In Bihar, Andhra Pradesh and Rajasthan. Felspar, Quartz and mica are the usual minerals.

4. **Lamprophyres.** These are dyke-rocks and are believed to have arisen from the same general body of magma. Mostly occur in the coal-fields where they are more prominent. In case of mica-lamprophyres—biotite is the dominant-mineral. In Hornblende lamprophyres, hornblende is dominant. In Augite lamprophyres—augite is dominant. But their occurrence is rare.

5. **Anorthosite.** In Eastern Ghats, they are found, in Bankura of West Bengal, Sitampundi Compbx of Tamilnadu, Banpur, Angul of Orissa. Dominantly composed of labradorite feldspar.

CHAPTER 52

ORIGIN OF SEDIMENTARY ROCKS

Sedimentary rocks are secondary rocks, which are constituted of sediments. The sediments are formed by the mechanical, or chemical activities of the natural agencies like running water, blowing wind, percolating water (underground), glaciers etc., which causes disintegration as well as decomposition of the pre-existing rocks. The products of decay are transported to some depositional sites by the natural agencies, where they get deposited and with subsequent compaction form sedimentary rocks. The secondary rocks usually accumulate under a great variety of physico-chemical condition and consequently show great variation in mineral, chemical composition as well as in texture. Weathering is the most important process that operate in the formation of sedimentary rocks.

Weathering takes place by three methods as

(i) Mechanical disintegration, associated with physical factors.

(ii) Rock decomposition, associated with chemical factors.

(iii) Biological weathering associated with the activities of organisms.

(i) **Mechanical disintegration.** It is due to frost action, thermal expansion and contraction, aided with gravitational forces. By such mechanical disintegration, reduction of size and disaggregation of rocks takes place.

(ii) **Chemical weathering.** The effectiveness of the chemical constituents of the atmosphere (as moisture, carbon dioxide, oxygen etc.) depends on the composition of the rock and the size of the particles that make up them. Important processes are oxidation, hydration, carbonation etc.

(iii) **Biological weathering.** It is due to the physical forces exerted by roots on the adjacent rocks and also due to chemical activity of its products of decay. The decaying products of plants and animals produce various types of acids and alkalies which play roles in the decomposition of rocks.

Products of weathering :

(a) The first product of weathering is a mantle of broken and decomposed material of varying thickness and composition called the regolith, which covers the areas except those from which it is removed as soon as formed.

(b) **Soluble salts.** These are carried away alongwith the transporting media in solutions.

(c) **Colloidal substance.** These are carried by ground water and stream. These are like $Al(OH)_3$, $Fe(OH)_3$ etc.

(d) **Insoluble products.** It includes clay minerals, quartz grains, undecomposed feldspar with some chemical resistant minerals like zircon, tourmaline etc., which are found at the site of weathering and later transported to the sites of deposition by geological agents.

Deposition. It may be mechanical or chemical, according to which mechanically transported material gets deposited first and the soluble materials get deposited through the process of chemical precipitation. Sometimes, the activity of bacteria is believed to be effective in the deposition of ferruginous soluble substances like bog-iron ore.

According to the environment of deposition the quality and quantity of the deposition is effected. Besides the assemblages of sedimentary rocks differ from environment to environment. While the continental environment includes-fluviatile, estuarine, lacustrine (fresh and salt water), deltaic, aeolian, glacial etc., the marine environment includes—shallow water, deep water abyssal etc. environment.

CHAPTER 53

CLASSIFICATION OF SEDIMENTARY ROCKS

Sedimentary rocks are due to the integration of the products of weathering; so any classification to be acceptable must take into account : (*a*) the genetic aspect of the sediments, (*b*) their textural characteristics, (*c*) their mineralogical composition as well as (*d*) their structural peculiarities. The present classification which is the most acceptable one has been proposed by E.W. Spencer, and the basis of classification is the mode of formation of the sediments.

Sometimes the products of weathering are carried down by the natural agencies and sometimes they are found at the place of their origin Accordingly there are two classes as

    1.   Residual deposits.   2.   Transported deposits.

1.   Residual deposits. These are also known as sedentary deposits. These are formed due to accumulation and consolidation of those materials which were left as residue during the operation of the weathering processes and transportation. These are the insoluble products of rock weathering which still mantle the rocks from which they have been derived.

They include the following rock types :

    (*i*) **Terra-rosa.** These are the insoluble residue of clay and other matter left behind after solution of limestone.

    (*ii*) **Laterite and bauxite.** In tropical and sub-tropical regions a reddish, porous and concretionary material is found to cover vast areas. They generally consist of a mixture of hydrated ferric-oxide with hydroxide of aluminium in varying proportions. These are called laterites.

When the aluminous content predominates the colour lightens to yellowish or whitish and the rocks become earthy or clay-like. It is called *bauxite*.

2.   Transported deposits. These are formed from the materials that have been transported both mechanically by traction, and suspension and chemically in solution. Besides, some organic processes also play active roles in the formation of transported deposits. The transported deposits are classified into two groups :

(a) Clastic rocks.   (b) Non-clastic rocks.

(a) **Clastic rocks.** These are detrital or fragmental rocks and are carried and deposited by mechanical means. On the basis of mode of transport and grain size, the clastic rocks are classified as follows :

(i) **Rudaceous rocks.** Very coarse grained rocks where the size of the grains are those of boulders. These are transported in traction, *i.e.*, by rolling or creeping. Also known as *rudites*.

e.g.,   (a) Conglomerates—fragments are rounded.

(b) Breccia—fragments are angular.

(ii) **Arenaceous rocks.** These rocks consist chiefly particles of sandgrade. They are transported in *saltation*. Also known as arenites, *e.g.*, sandstone, arkose, graywacke, grits etc.

(iii) **Silt rocks.** Here the constituent particles are finer than common sand and coarser than clay. They are transported by suspension, *e.g*, loess.

(iv) **Argillaceous rocks.** These are made up of clay particles, usually transported in suspension. *e.g*, Clay, Mudstone, Shale etc.

(b) **Non-clastic rocks.** These are formed due to chemical precipitation as well as by biological means. Thus they are of two types :

(i) Chemically deposited sediments, and

(ii) Organic sediments.

(i) **Chemical deposition** :

(1) *Evaporites.* It is only due to evaporation and the deposits are like salt and gypsum.

(2) Through reaction between the components carried in solution ; siliceous, calcareous, ferruginous and carbonates deposits are produced in this way. The examples are

1. **Siliceous deposits.** Chert, flint, siliceous sinter etc.

2. **Calcareous and carbonates**—limestone, dolomite, calc-sinter or travertine.

3. **Ferruginous deposits.** Iron salts, hematite, goethite siderite etc.

(ii) **Deposits of organic origin.** These are the products of accumulation of organic matter preserved under suitable conditions.

A rock of organic origin may be built up directly from the beginning as a quite solid material as in the case of coral rocks and algal limestones. In other cases the deposition may be bio-chemical or bio-mechanical. They are mainly of five types :

1. Siliceous. Radiolarian ooze ; diatoms are lowly plant organisms which secrete silica.

2. Calcareous. Due to biomechanical processes as well as biochemical processes these deposits are formed. Fossiliferous limestone, chalk, marl etc. are the examples.

3. Phosphatic. As calcium phosphate is utilized by certain organisms, especially fish and brachiopoda, the remains of these organisms accumulate on the sea-floor forming phosphatic deposits.

'Guano' is directly of organic origin.

4. Ferruginous. By the activities of bacteria, *e.g.*, Bog-iron ore.

5. Carbonaceous. Coal formations.

Besides the above, the consolidated pyroclastic rocks are also considered as sedimentary rocks.

CHAPTER 54

STRUCTURES OF SEDIMENTARY ROCKS

Sedimentary rocks are the secondary rocks, in which the presence of different layers, beds or strata distinguishes them from the igneous and metamorphic rocks. Sedimentary structures are both organic and inorganic in origin. Depending on the mechanism of formation, the inorganic structures are classified into

I. Primary.   II. Secondary structures

I. Primary structures. The primary structures are due to mechanical action of current and show the following characteristics :

    (*a*) these structures show palaeo-current condition,

    (*b*) rate of supply of sediments,

    (*c*) mode of transportation,

    (*d*) environment of deposition, and

    (*e*) top and bottom of beds etc.

The primary sedimentary structures include the following :

    (*i*) **Bedding or stratification**. Insoluble mechanically transported material is deposited in layers on the surface of accumulation which may be horizontal or inclined. Stratification may be the result of variations in composition of different layers, colour of layers, textures of the layers and porosity of the layers. These are most conspicuous particularly in the rocks formed under water. Glacial, chemical and biological deposits lacks stratification. Aeolian deposits seldom show bedding.

If the individual layers are extremely thin the structure is known as *lamination* and the layers as *laminae*. There are two types of laminations :

    1. **Dimictic lamination.** Where the contact between two laminae is sharp.

    2. **Symictic lamination.** Here the contact between the laminae is rough.

The plane of contact is known as laminating plane, *e.g.*, varve.

When the bedding planes are disposed approximatlely parallel to one another, they are known as *concordant*. If bedding planes are inclined to the major lines of stratification, they are said to be discordant.

(*ii*) **Cross-stratification.** Here the beds are found to lie slightly oblique to the major bedding planes and bound by layers of concordant bedding. Mostly found in arenaceous rocks.

It is formed due to change in the velocity and direction of flow of streams. The oblique lines of a cross-bedded layer always meet the upper concordant bedding at a higher angle and lower portion tangentially.

In wind formed current beddings, the laminations are curved and of larger magnitude. The cross-stratification is also known as current bedding or false bedding etc. When the top and bottom beds converge at a point, it is said to be wedge-cross stratification.

(*iii*) **Torrential bedding.** It shows an alternation of coarse current bedded material and finer horizontal laminae. Here the foreset beds are straight and they characteristically develop in alluvial fans.

(*iv*) **Graded bedding.** In this case there is a gradation of grain size from coarser at the bottom to finer at the top. It is having a sharp contact with the underlying strata. This in consolidated form is known as turbidites. Graded beddings are seen in 'Graywackes.'

(*v*) **Ripple marks.** These are minute undulations formed due to current or wave action, developed on arenaceous rocks. These are of two types :

1. Oscillation or wave ripple 2. Current ripple.

1. **Oscillation ripples.** These are symmetrical with sharp crests and broad rounded troughs. These are shallow water features and are indicative of a stagnant body of water frequently agitated by waves.

2. **Current ripples.** These are asymmetric in nature, having parallel, long, more or less equidistant ridges trending in straight lines at right angles to the direction of current. Here both the crests and troughs are rounded.

Aqueous ripples contain finer materials at the crest and coarser material at the troughs but in the case of aeolian ripples coarser materials are found at crests. Aeolian ripples are invariably asymmetric in nature.

(*iv*) **Mud-cracks.** These cracks typically develop in clayey sediments due to prolonged exposure to the atmosphere. These are also known as shrinkage cracks or suncracks. They are wider at the top but tapers towards the bottom.

(*vii*) **Rain-prints.** These are shallow depressions surrounded by a low-ridge formed by the impact of the drop, hailstones, bubbles etc.

(*viii*) **Imbricate structure.** In conglomerate and pebble beds fragments having a noticeable elongation are sometimes deposited with the long axes of the pebbles lying more or less parallel to one another, leaning in the direction of current flow.

(*ix*) **Tracks and trails.** These are markings indicative of the some animal over soft sediments. Trails are the winding passages through which long bodied and short bodied animals moved.

(*x*) **Convolute bedding.** Here, the sedimentary lamina are contorted into a series of anticlines separated by broad synclines. Distortion increases upwards but it is confined to one bed and is often abruptly truncated by overlying sediments.

II. Secondary structures. These are the products of chemical action contemporaneous with sedimentation or shortly thereafter and includes :

(*a*) **Concretions.** They are spherical to elliptical bodies, usually small and of diverse chemical nature than the rocks in which they occur. They include nodules, oolites, pisolites, geodes etc.

(*b*) **Solution structures.** Irregular openings commonly in calcareous rocks and are produced due to ground water action *e.g.*, Vugs.

Organic structures. Fossils impressions, as well as petrified remains of animals or plants are the common organic structures found in sedimentary rocks.

Sole structures. These are the structures preserved on the base of a bed which is sharply differentiated lithologically from the bed below, *e.g.*, sandstone overlying a shale. They also include flute cast (which are grooves eroded by turbulent flow and later filled with coarse sediments) and groove casts formed on the surface of sandstone layers, tracks and trails, prod marks, etc.

CHAPTER 55

RESIDUAL DEPOSITS

Residue left as *insiiu* after weathering followed by transportation gives rise to residual deposits in due course. These are the insoluble products of rock weathering, which still mantle the original rocks from which they have been derived.

The residual deposits are formed because of the accumulation of insoluble products of rock weathering, when the other constituents are removed by the transporting agency. As a result there is a decrease in volume, affected almost entirely by surficial chemical weathering.

Mode of formation. For the formation of residual deposits, the following conditions are required :

(*i*) Humid-tropical climate.

(*ii*) Presence of rocks, where some of the constituents are soluble and a majority of constituents should be insoluble.

(*iii*) The relief must not be very great.

(*iv*) Long continued crustal-stability is essential.

(*v*) There should be conditions for good drainage.

Under conditions of poor drainage clay minerals like *kaolinite*, illite, montmorillonite etc. are formed. If there are well defined wet and dry seasons and fairly good drainage, the clay minerals are decomposed to form laterite. But with an evenly distributed rainfall through out the year, iron-oxide tends to be separated in the solution because of aeration leaving behind mainly aluminium hydroxide, which is called 'bauxite'.

Thus the examples of residual deposits are clay, laterite and bauxite.

Besides, Terra Rossa forms a good example of residual deposits, which is the insoluble residue of clay and other mineral matter left behind after solution of the limestone.

Residual deposits sometimes contain valuable ore deposits and the process is termed as residual concentration.

CHAPTER 56

TEXTURE OF SEDIMENTARY ROCKS

The word 'texture' refers to the size, shape, packing and fabric of the components of the rock. Since the sedimentary rocks are broadly classified as (1) exogenetic or clastic rocks and (2) endogenetic rocks or the chemically precipitated amorphous or crystalline rocks, accordingly their texture are also classified into two broad categories.

1. Clastic texture. It includes elements like :

(I) Size, (II) Shape, (III) Sphericity, (IV) Packing, (V) Fabric.

(*a*) **Size.** The grain size is dependent on the (*i*) mode of weathering, (*ii*) nature of the source rock, and (*iii*) kind and distance of transport and the nature of deposition.

Broadly, the size character of the sediments are described as either coarse, medium or fine. The size grades of the clastic particles, in the wentworth scale are indicated as follows :

Size	Name	Equivalent rocks	
(*i*) >256 mm in diameter	Boulders	Boulder	Rudaceous Rocks
(*ii*) 64 to 256 mm	Cobble	Cobble	Rudaceous Rocks
(*iii*) 4 to 64 mm	Pebble	Pebble	Rudaceous Rocks
(*iv*) 2 to 4 mm	Granule	Granule Stone	Arenaceous Rocks
(*v*) 2 to $\frac{1}{16}$ mm	Sand	Sandstone	Arenaceous Rocks
(*vi*) $\frac{1}{16}$ mm to $\frac{1}{256}$ mm	Silt	Siltstone	Argillaceous Rocks
(*vii*) $\frac{1}{256}$ mm or less	Clay	Claystone	Argillaceous Rocks

The size analysis results are represented in form of tables. histograms or frequency curves as well as by statistical methods. In statistical methods the following measurements are made :

(*a*) **Measurement of central tendency.** Which determines the average size of the distribution and refers to the overall competency of the transporting medium.

(*b*) **Measurement of dispersion.** It determines the turbulance of the transporting medium or the amount of reworking of the sediment has undergone prior to final burial.

(*c*) **Measurement of skewness.** It determines whether the coarser and finer admixture occur in same proportion, in a sediment.

(*d*) **Measurement of kurtosis.** It determines the peakedness or flat-toppedness of a distribution.

Geological Significance. The size analysis indicates the following :

1. Provenance :

(*i*) Composition of the source rock is an important factor that determines the extent to which the component minerals are susceptible to weathering and liable to pass on to the products and reduced in size and shape.

(*ii*) Besides the coarse or fine texture is also a function of the source area.

2. Transport. As we know, more the distance of transport, finer is the grain size. Besides, the character of the sediments are also governed by the mode of transport, *i.e.*, traction, saltation, suspension, which is a function of the kinetic energy of the transporting medium. Higher the energy, the coarser particles can be transported.

3. Depositional environment.

4. Palaeo current. The coarser sediments carried in rolling are deposited in basin margin, whereas the finer sediments are gradually carried to the centre. As such there is a regular variation in grain size from the margin to the centre of the basin.

5. Transporting medium Graded sediments are the result of long continued transport, while ill-sorted sediments of a rapid and confused deposition like glacial deposits. Aeolian deposits are apt to be well graded and uniform.

6. Tectonics of depositional site. With an increase in the rate of subsidence of the area of deposition, the grainsize decreases and the average sorting is poor. Under stable condition, the grainsize is determined by the texture of the material available for reworking and sorting steadily improves.

II. Shape. It is defined as the sharpness of corners and edges of a clastic fragment. Accordingly the shape may be angular, sub-angular, sub-rounded, rounded, well rounded, etc. The shape of the sedimentary grains is determined by :

(*i*) Original shape of the mineral, (*ii*) Stability of minerals, and (*iii*) amount and nature of transport.

III. Sphericity. It is defined as the extent to which a particle approaches a sphere. It depends on (*i*) distance of transport, (*ii*) mode of transport, and (*iii*) provenance.

(*i*) Longer the distance of transportation, more chances of being reworked and therefore more the degree of round-ness. Besides, wind produces perfect rounding, glacier does the least.

(*ii*) The mode of transport like traction, saltation and suspension produces particles of variable roundness.

(*iii*) Generally the elongation quotient is maximum in meta-morphic rocks, less in igneous rocks and very less in sedimentary rocks. Thus when the source rock is of metamorphic origin, the sphericity of the clastic grains is not that much pronounced.

IV. Packing. It is the manner of aggregation of sedimentary grains, which are held together in place in the earth's gravitational field. There are six methods of packing out of which the rhombohedral packing is the most compact and tight whereas the cubic packing is the loosest possible packing It determines the porosity and permeability of sedimentary rocks.

V. Fabric. It is the arrangement of the clastic particles in sediments. It is defined as the orientation of the grains or lack of it with which the sedimentary rock is composed. Pebbles, sand grains, mica-flakes etc. are the most useful fabric elements, also some fossils like gastroped shells etc. It determines the palaeo-current direction.

2. Non-clastic textures. It is formed as a result of deposition through chemical reaction. They are transported chemically by getting dissolved in the transported media but reappear due to precipitation or evaporation. It is of two types :

(*a*) Crystalline texture.
(*b*) Non-crystalline texture.

(*a*) **Crystalline texture.** They are formed due to direct precipitation from a saturated solution, and the result is an interlocking aggregate of crystals.

(*b*) **Non-crystalline texture.** When colloids coagulate they form a gelatin like mass. This gelatinous mass may lose some of the water in it and eventually harden to form an amorphous mass. Nodular, oolitic, spherulitic textures are the examples. Many concretionary, botryoidal, reniform, nodular, oolitic and pisolitic textures are believed to be of colloidal origin and they show non-crystalline textures as described above.

CHAPTER 57

MINERALOGICAL CHARACTER OF CLASTIC SEDIMENTARY ROCKS

It is well known that minerals are in equilibrium with the environment in which they are formed and hence stable. Some are stable over a wide range of temperature-pressure conditions while others yield rather readily. Different silicate minerals of which most of the rocks are composed have different stability. The stability of minerals is approximately in the reverse order to that of their crystallisation from a magma. It is therefore, commonly observed that rocks of basic and magnesium varieties breakdown much faster than the acid and ferruginous ones.

The mineralogical character of the sedimentary rocks are therefore mostly governed by the stability-characteristic of the minerals.

The most stable minerals, which are found quite abundantly in the sedimentary rocks are quartz followed by muscovite.

The various feldspar, alongwith zircon rutile, tourmaline, monazite, garnet etc. are fairly stable.

Biotite, hornblende, apatite, ilmenite, magnetite, staurolite, kyanite, topoz, sphene etc. are minerals of the 'least-stable' category.

Minerals like augite, hypersthene, diopside, actinolite, olivine etc. are highly unstable.

In view of the stability characters of the minerals, in most of the sedimentary rocks we find the minerals like quartz and muscovite whereas the presence of unstable minerals like augite, hypersthene etc. is rarely found.

In general, the sandy rocks consists largely of stabler primary minerals, while the fine grained clayey materials are constituted mostly by the ferromagnesian silicates and feldspars.

CHAPTER 58

SAND STONES

Sandstones belong to the arenaceous group of sedimentary rocks. They consist of detrital particles, usually but not necessarily rounded; of sand-grade (*i.e.*, the size range is 1/16 to 2 mm) held together with a matrix or cement. The mineral composition of an 'average sandstone' is as follows :

Quartz 68%   Feldspar 12%

Mica 6%   Carbonates 11%

Miscellaneous 4%.

The classification of sandstones proposed by various geologists like Pettijohn, Krynine, Dott etc. indicates that chiefly there are three major groups of sandstone as

1. Arkose.   2. Graywacke.   3. Quartz-arenite.

1. Arkose. The important characteristics of Arkose are as follows :

(*a*) **Mineral composition.** It is composed primarily of feldspar and quartz. Generally quartz exceeds feldspar. The feldspar is largely potash feldspar. Other constituents form 5 to 15% of the rock and are mainly micas. The composition of an arkose corresponds to that of granitic rocks from which it is presumably derived.

(*b*) They are typically light pink or light gray in colour.

(*c*) Arkose commonly shows current-bedding and are terrestrial-shallow water deposit.

(*d*) The cementing material is chemical and the detrital matrix is less than 15% or absent.

(*e*) It is a product of oxidising environment.

(*f*) When feldspar exceeds quartz in an arkose it is termed as feldspathic arkose.

(*g*) Of all the sandstones, arkose constitute only about 15%.

2. Graywacke. They show the following important characteristics :

(*a*) **Mineral composition.** These rocks are composed mainly of rock-fragments. In this case feldspar exceeds quartz.

(*b*) Graywacke is dark in colour.

(*c*) It commonly shows.graded bedding and are mostly submarine, deep water deposit.

(*d*) These are the sandstones with high detrital matrix content and no chemical cement.

(*e*) It is a product of reducing environment.

(*f*) The matrix has the composition of slate and imparts both the distinctive colour and general toughness to the rock. Like slate, the matrix consists of a fine grained mixture of white mica, chlorite and quartz.

(*g*) Of all the sandstones, graywackes constitute about 50%.

(*h*) The composition of graywackes corresponds to the basic igneous rocks which are the source-rocks.

3. Quartz arenite. These are also known as orthoquartzites, which are pure sandstones and show the following characteristics :

(*a*) They are composed of quartz to the extent of 95% or more.

(*b*) They are presumed to be associated with the peneplanation stage of geomorphic cycles.

(*c*) In this case the detrital matrix is absent or less than 15% at the most.

(*d*) Here the feldspar as well as the rock-fragments constitute even less than 5%.

(*e*) They may be derived from the lithic sandstones or from arkosic sandstones. In the former case they are charactefised by metamorphic quartz and chert particles ; whereas in the latter case the chert is absent or nearly so (under 5%) and the quartz is igneous-quartz.

CHAPTER 59

SILICEOUS DEPOSITS

These are deposits formed through solution by two distinct processes *viz*.

(1) Chemical.   (2) Organic.

1. Siliceous deposits of chemical origin. Quartz is insoluble but other forms of a silica are fairly soluble in natural alkaline waters but essentially nearly all silica is transported in colloidal rather than true solution.

Siliceous sinter. In volcanic region, sometimes hot springs bring up silica and deposit it in the mounds and terraces about the orifices of eruption. This material consists of cryptocrystalline or opaline silica. The deposition is mainly due to evaporation and cooling of waters but certain algae and organisms also initiate precipitation of silica.

Chert and flint. They occur as irregular nodules or tabular masses, which consist of minutely—crystalline or cryptocrystalline silica. Most cherts were deposited directly from the sea water replacing carbonate rocks with which they are closely associated and frequently interbedded.

2. Organic siliceous deposit. They result from direct precipitation by organisms which extract silica as opal from natural waters and bodily contribute to the siliceous deposit. Radiolarians, diatoms, algae, sponges are the organisms involved in the deposition of silica.

(*a*) **Radiolarian oozes.** These are a group of single-celled animals which construct skeletons of silica, which owing to their relative insolubility can sink to greater depths and form radiolarian ooze. The radiolarian ooze are closely associated with the 'Abyssal-red clay'.

(*b*) Diatoms are microscopic plants, which secret minute, ornamented, spherical and discoidal body composed of silica and which ultimately gives rise to siliceous deposits after their death.

CALCAREOUS DEPOSITS

Calcareous deposits may be of chemical as well as of organic origin.

1. Chemical or inorganic-limestone. They result by precipitation through evaporation, chemical reaction or by other physicochemical means.

(*a*) **Travertine or tufa.** These are limestones formed by the evaporation of spring and stream waters containing calcium carbonate in solution.

(*b*) **Kankar.** It is a nodular iron-rich calc-sinter, formed by capillary action.

(*c*) **Stalactite and stalagmite** (dripstone).

(*d*) **Oolitic limestone.**

2. Organic-calcareous deposits :

(*a*) **Fossiliferous limestone.** These are formed from the shells of marine animals and also by corals.

(*b*) **Chalk.** It is a fossiliferous limestone composed of the shells of protozoans (single-celled animals, *i.e.*, the foraminifera etc.)

(*c*) **Marl.** This name applies to mixtures of shells and shell fragments with muds and sand. Thus it is an impure limestone.

CHAPTER 60

METAMORPHISM

Metamorphism is the mineralogical and structural adjustment of solid rocks to physical and chemical conditions which have been imposed at depths below the surface zones of weathering and cementation, and which differ from the conditions under which the rocks in question originated. Thus metamorphism is the response of the solid rocks to pronounced changes of temperature, pressure and chemical environment.

Metamorphism stands midway between diagenesis and general melting of rocks. Important features of metamorphic changes :

(*i*) The bulk chemical composition of the metamorphic rock is the same as that of the rock from which it is formed. Thus metamorphic changes are isochemical changes.

(*ii*) The structural and textural characteristics of the metamorphic rocks are the outcome of the structure and texure of the pre-existing rocks and temperature-pressure condition of the metamorphic changes.

(*iii*) The changes in metamorphism takes place in an essentially solid medium.

Metamorphic rocks which are derived from igneous rocks are known as Orthometamorphic rock, and those which are derived from the sedimentary rocks, are known as parametamorphic rocks.

Agents of metamorphism. The agents which are mostly responsible for bringing about metamorphic changes are as follows :

1. Temperature.
2. Pressure :
   (*a*) hydrostatic or uniform pressure,
   (*b*) directed pressure or stress.
3. Chemically active fluids.

1. Temperature. It may be supplied by geothermal gradient, magmatic heat, frictional heat and by radioactive disintegration. The temperature range within which metamorphic changes take place is from 200° to 700°C. However, in certain cases a temperature of 1000° to 1200°C may be encountered.

Temperature accelerates the processes of reaction, increases the volume of the rocks, remove volatiles and moisture contents of the rocks.

The following types of metamorphism are said to be the result of temperatures effect on rocks :

(*i*) **Pyrometamorphism.** At 800° to 1000°C, in the immediate viciniy of the intrusives.

The induration, backing, burning and fritting effects of lava flows and intrusions on neighbouring rocks is known as Caustic metamorphism or optalic metamorphism.

(*ii*) **Contact metamorphism.** It occurs around larger intrusives at comparatively low-temperature. It includes :

(*a*) *Normal contact matamorphism.* Where rocks are simply crystallized without new mineral formation.

(*b*) *Pneumatolytic, additive or metasomatic.* The composition of the rocks is vastly modified depending on the addition of material from magmatic emanations.

(*c*) *Injection metamorphism.* Here with the intrusion of magma or its residual liquid may alter the intruded rock substantially.

(*iii*) **Auto-metamorphism.** It is the mineralogic readjustment of an igneous assemblage to the falling temperature as the body of the igneous rock cools. It includes uralitisation, serpentinisation etc.

(*iv*) **Retrograde metamorphism.** Also known as *diaphthoresis*, where mineralogical rearrangement of high temperature assen.blage to a low temperature one takes place.

2. (*a*) **Uniform pressure.** It is the hydrostatic pressure which increases with depth. Uniform pressure and temperature can both dominate together at great depths. There is a reduction in the volume of the rock and a change in the mineralogical composition. It is known as *Plutonic Metamorphism*, e.g., Granulites, Eclogites.

Load metamorphism. It is due to the vertically acting stress of superincumbent rock masses aided by high temperature.

(*b*) **Directed pressure.** It is produced mostly by orogenic movements. It dominates at or near the surface. It results in crushing and granulation of minerals, without the formation of any new mineral. It is also known as dynamic metamorphism or cataclastic metamorphism, e.g., Mylonites.

Where both heat and stress **dominates**, the metamorphism is known as Regional Metamorphism. Here the country rocks are subjected to changes both in mineral composition and texture. It is also known as dynamothermal metanorphism.

3. Chemically active fluids. These are from the following sources :

(*a*) Meteoric water.   (*b*) Juvenile water.

Water carries minerals in some cases in solution and also serves as a medium in which chemical changes occur with ease. Chemical activity is more pronounced in the vicinity of the igneous intrusions.

GRADES OF METAMORPHISM

The degrees of metamorphism or grades **depend** upon the extent to which the agents were in operation during the process. According to the temperature, pressure condition, there are usually three grades of metamorphism and accordingly there are three zones :

1 Epizone. It is the zone of low-grade metamorphism, where temperature ranges from 100 to 300°C, pressure is low to moderate. It]is charcterised by the presence of the hydrous minerals.

Sericite, muscovite, chlorite, biotite, talc, actinolite, epidote, andalusite etc.

Rocks. Slates, phyllites, chlorite-schists, muscovite-schists, biotite-schists.

2. Mesozone. This is the zone of medium grade metamorphism, where the temperature ranges from 300 to 500°C, pressure is moderately high. It occurs at an intermediate depth, *i.e.,* between 5 to 10 miles.

Minerals. Biotite, andalusite, cordierite, quartz, hypersthene, almandine, orthoclase, ilmenite etc.

Rocks. Phyllites and mica-schists.

3. Katazone. It is the zone of high grade metamorphism, where the temperature ranges from 500 to 650°C, pressure is high. It occurs at a depth of 9 to 13 miles. It is characterised by anhydrous and antistress minerals.

Minerals. Biotite, alkalifeldspar, plagioclase, quartz, garnet, silimanite, kyanite, etc.

Rocks. Gneisses of various types, hornfels etc.

In case of Regional metamorphism, the following grades have been identified :

1. Zone of chlorite.   2. Zone of biotite.
3. Zone of garnet.   4. Zone of staurolite.
5. Zone of kyanite.   6. Zone of sillimanite.

These zones are according to the progressive grade of regional metrmorphism.

CHAPTER 61

METAMORPHIC STRUCTURE

The metamorphic structures are determined by definite mechanical conditions and also by recrystallisation. Relict (remnant of the original structure) and crystalloblastic (metamorphic crystalline structure) structures may exist side by side. The relict structure may be used to trace back the nature of the original rock and the magnitude of alteration it has undergone. Five major types of metamorphic structures have been recognised as follows :

(*i*) Cataclastic texture.   (*ii*) Maculose structure.

(*iii*) Schistose structure.   (*iv*) Granulose structure.

(*v*) Gneissose structure.

Metamorphic rocks derived from the sedimentary rocks are known as parametamorphic rocks. Those which are derived from gravel rocks are known as 'psephitic rock', from aranaceous rocks are called psammitic rocks and those from argillaceous ones are said to be 'pelitic rocks.'

(*i*) **Cataclastic texture.** It is produced under stress and in absence of high temperature, whereby rocks are subjected to shearing and fragmentation. Only the durable mineral partly survive the crushing force and the less durable ones are powdered. Thus, when resistant minerals and rock fragments stand out in a pseudo porphyritic manner in the finer materials, it is known as 'porphyroclastic structure.' Phenocrysts are called 'porphyroclasts' Argillaceous rocks develop slaty cleavage, harder rocks may be shattered and crushed forming crush breccia and crush conglomerate. When the rocks are highly crushed into fine grained rocks, they are known as mylonites Since these structures are formed due to cataclasis, they are, as a whole, known as cataclastic structure.

(*ii*) **Maculose structure.** It is produced by thermal metamorphism of argillaceous rocks like shales. Here, larger crystals of andalusite, cordierite and biotite are sometimes well developed giving a spotted appearance to the rocks. The well developed crystals are known as 'porphyroblasts' with increasing degree of metamorphism, the spotted slates pass into extremely fine grained granular rock known as Hornfels.

(*iii*) **Schistose structure.** Here the platy or flaky minerals like the micas and other inequidimensional minerals show a preferred orientation along parallel planes, under the effect of the stress dominating during metamorphism. The longer directions are parallel to the direction of maximum stress. Schistosity is the property or tendency of a foliated rock, whereby it can be readily split along foliation plane.

(*iv*) **Granulose structure.** This is found in the rocks composed of equidimensional minerals like quartz, feldspar and pyroxenes. They are formed by the recrystallisation of pre-existing rocks, under uniform pressure and great heat. The typical texture is coarsely granoblastic. These structures are also known as '*sacchroidal*. Quartzites and marbles are typical examples of this structure.

(*v*) **Gneissose structure.** It is a banded structure due to alternation of schistose (dark coloured) and granulose (light coloured) bands and is produced by highest grade of metamorphism, typically by regional metamorphism. The bands differ from one another in colour, texture and mineral composition. Gneisses typically show this type of structure, hence the name.

CHAPTER 62

METAMORPHIC TEXTURE

The texture of the original rock which has undergone metamorphism is sometimes found to exist in the metamorphic rocks. Such textures are called '*relict or palimpset texture*.' In describing metamorphic textures the terms '*blastic*' or '*blast*' are used as a suffix to represent the metamorphic equivalents of igneous textures of similar look.

Recrystallisation of minerals produces a '*crystalloblastic texture*', which is similar to the holocrystalline texture of igneous rock. If during metamorphism a texture similar to porphyritic comes into existence, the same is described as '*porphyroblastic texture*'.

In case of palimpset textures, if for example a porphyritic igneous rock is metamorphosed and the original texture continues to occur in metamorphosed ones, the resulting texture will be said to be '*blasto-porphyritic*'. Thus, palimpset textures are prefixed by '*blasto*'.

Similar to the igneous texture where the minerals have perfect crystal outlines, such grains are called '*idioblastic*', if not 'xenoblastic'. Where the recrystallised mineral grains are found to be equidimensional, the texture is said to be '*granoblastic*'. '*Helicitic texture*' is a term commonly applied to 'S' shaped or 'Z' shaped trails of inclusion in poikiloblastic crystals, especially garnets and staurolites found in regionally metamorphosed rocks.

MINERALOGICAL COMPOSITION OF METAMORPHIC ROCKS

(*i*) **Stress and anti stress minerals.** Stress minerals are produced as a result of stress and have a stable existence only under stressed conditions. Kyanite, garnet, chloritoid, staurolite, epidote, zoisite, glaucophane, anthophyllite etc. are common stress minerals.

The anti-stress minerals are those which are formed conveniently under uniform pressure. These minerals are incapable of withstanding high shearing stresses. Such minerals therefore do not occur in highly deformed rocks. They may include—sillimanite, cordierite, anorthite, felspathoids, andalusites, alkali feldspars etc.

Antistress minerals are of low-density while stress minerals in general are dense.

(*ii*) The follo⋏ing are the typical metamorpic minerals :

Aluminosilicates like andalusite, kyanite, sillimanite, staurolite, cordierite, epidote, tourmaline, talc, chlorite, zeolites, graphite, pyrite and pyrrohtite etc.

METAMORPHIC CLASSIFICATION

Metamorphic rocks have been classified on the basis of several factors like :

1. The parent rocks from which they have been metamorphosed.
2. Structure, texture and predominance of agents.
3. Mineralogical assemblages etc.

However, the first and the last factors are quite significant in the classification of metamorphic rocks.

In case the parent rocks are of igneous origin which have subsequently undergone some or other metamorphic changes, the resulting rocks are known as

(*a*) *Orthometamorphic* rocks or Meta-igneous rocks.

When sedimentary *rocks* undergo metamorphic changes, the resulting rocks are said to be (*b*) *Para-metamorphic* rocks or Metasedimentary rocks.

On the basis of the mineralogical assemblages also attempts have been made to classify the metamorphic rocks. As we know, whenever metamorphism is ideally complete, the product is an assemblage of minerals in chemical equilibrium with one another and in most metamorphic rocks, this ideal condition appears to be at least closely approached. The actual mineral composition of a metamorphic rock is determined by two factors :

(*i*) The initial composition of the rock and the extent to which material have been added or subtracted during metamorphism.

(*ii*) The degree of metamorphism.

Accordingly the facies concept also came into being in metamorphism. It has also been observed that the number of mineral assemblages produced by metamorphism is between two and six. On the basis of type of metamorphism, *i.e.*, wheth⣇r contact metamorphism or regional metamorphism as well as the range of temperature the following classifications have been made :

   1.   **Contact metamorphism :**

      (*i*) Albite-epidote-hornfels facies.

      (*ii*) Hornblende-hornfels facies.

      (*iii*) Pyroxene-hornfels facies.

      (*iv*) Sanidinite facies.

   These four facies have been distinguished in the ascending order of temperature of formation and have characteristic mineral assemblages.

   2.   **Regional metamorphism.**   There are six facies included in this group in ascending order of temperature of formation, as

      (*i*) Zeolite facies.   (*ii*) Green-schist facies.

      (*iii*) Glaucophane-schist facies.   (*iv*) Amphibolite facies.

      (*v*) Granulite facies.   (*vi*) Eclogite facies.

   All the possible mineralogical assemblages have been represented by the above facies types. Accordingly metamorphic rocks are classified.

   Besides the above, some geologists classify metamorphic rocks into two broad categories :

      (*a*)   Foliated, and (*b*)   Non-foliated.

   (*a*)   **Foliated rocks.**   These are characterised by parallel arrangement of slaty minerals, such as the micas. Foliation is produced during regional metamorphism. The degree of foliation is related to the intensity of metamorphism. The foliations are ultimately changed into bands.

   The most common examples are slates, schists and gneisses.

   (*b*)   **Non-foliated metamorphic rocks.**   In these rocks the mineral grains are equidimensional, hence there is no specific orientation. Mostly they are the products of thermal metamorphism or contact metamorphism.

   Marble, quartzite and hornfels are some of the common examples of this type.

CHAPTER 63

PETROGRAPHIC CHARACTERISTICS OF METAMORPHIC ROCKS

Quartzite. Quartzite is the metamorphic equivalent of quartz sandstones. Sometimes the boundaries of the original grains of sand are visible but they have become firmly cemented together. Thus quartzite is a light coloured, medium specific gravity, showing typical vitreous lusture and conchoidal fracture and sacchraoidal structure. The principal constituent is quartz and this is a non-foliated metamorphic rock. It is produced by thermal or regional metamorphism of arenaceous rocks such as sands, sandstone and quartz-veins.

Slate. These are the most prefectly foliated metamorphic rocks. These are gray to black, fine grained rocks composed of finely divided micaceous minerals as micas, chlorite etc. with minor quantities of quartz, feldspar etc. These are the metamorphic equivalents of shales and mudstones. Carbonaceous matter is usually found associated with slates. The rock is characterised by the development of extremely good close-spaced cleavage planes, which are called 'slaty cleavages'.

Schist. These are strongly foliated rocks of medium to coarse crystalline texture. Foliation or schistosity is caused by parallel or nearly parallel alignment of micaceous minerals. The most common minerals in schists are quartz, feldspar and micas. These are produced by medium to high-grade of regional metamorphism of argillaceous or quartzo-argillaceous sediments.

Gneiss It is a typical foliated metamorphic rock usually having alternate dark and light streaks or bands. The light coloured bands consist predominantly of light coloured minerals as quartz, feldspar etc., whereas the dark bands consist of micaceous and/or the accicular ferromagnesian minerals. These rocks represent highest grade of regional metamorphism of quartz-felspathic rocks (granites) or quasi-argillaceous sediments.

Marble. This is a non-foliated metamorphic rock, which is the metamorphic equivalent of calcite, limestone or dolomite. It is a soft, compact, sacchraoidal rock. The texture is massive granular.

Hornfels. It is a hard, compact, fine grained rock composed principally of quartz, micas, iron-oxide etc., in which the typical parallel texture of regionally metamorphosed crystalline schists is absent. These are non-foliated, dense and usually dark coloured metamorphic rocks.

PART VI

ECONOMIC GEOLOGY

CHAPTER 64

IMPORTANT TERMINOLOGIES IN ECONOMIC GEOLOGY

Ore-deposit. If a mineral-deposit is of sufficient concentration to be profitably worked, it is called an ore deposit.

Ore. An ore is a term applied to that part of a metalliferous mineral deposit, which can be used for profitable extraction of one or more metals. The economic minerals of an ore is called '*ore-minerals*'.

Gangue. The term is used to indicate the useless material assoicated with ore-minerals. The usual gangue minerals are quartz and other forms of silica, calcite, dolomite, siderite, barite, felds-pars, garnet, chlorite, fluorite, apatite, pyrite etc. and sometimes the gangue material is the country-rock itself in which the ore minerals occur.

Tenor. The metal content of an ore is called the tenor of the ore. It is generally expressed in percentage of the metal.

CHAPTER 65

OUTLINES OF PROCESSES OF FORMATION OF ORE-DEPOSITS

As we know, an ore is composed of ore minerals and gangue, which can be utilised for a profitable extraction of one or more metallic compounds or metals. The entire crust of the earth consists of minerals. They occur as solid masses, or rocks of which the earth's crust is composed, or as local accumulations of varying size, such as veins, pockets or impregnations in rocks.

The processes of formation of mineral deposits are grouped into three main types :

(A) Magmatic.   (B) Sedimentary.

(C) Metamorphic.

Each type of these processes includes a number of subsidiary processes associated with them. Mineral deposits formed due to the verious processes associated with magmatic activities are called *'Primary-Mineral Deposits'*. These are also called *'hypogene-deposits'*. Mineral deposits arising out of the processes of weathering, and activities of several geological agents are called *'Secondary Mineral Deposits'*. These are closely associated with the sedimentary processes of formation. Metamorphic mineral deposits are the outcome of metamorphic processes acting upon an earlier formed mineral deposits or rocks.

(A) The Magmatic process of formation of mineral deposits include the following processes :

1. Magmatic concentration.
2. Pegmatitic (pneumatolytic).
3. Contact-metasomatic process.
4. Hydrothermal processes.
5. Sublimation.

1. Magmatic concentration. As we know, magma consists of a multitude of constituents, which are in mutual solution. As the magma approaches the earth's surface its temperature and the external pressure drop, with the result of crystallization and differentiation of minerals in a definite sequence. The formation temperature of different magmatic deposits varies from 1500°C to 300C°.

The magmatic deposit are classified into two major groups, *viz.*, (*i*) Early magmatic deposit, and (*i*) Late magmatic deposit. The early magmatic deposits are believed to have been formed simultaneously with the host-rock, whereas the late magmatic deposits are formed towards the close of the magmatic deposits.

The early magmatic deposits are usually formed by

(*a*) Simple crystallization without concentration.

(*b*) Segregation of early formed crystals.

(*c*) Injection of materials, concentrated elsewhere by differentiation.

(*a*) **Dissemination.** It involves simple crystallization whereby early formed crystals are found disseminated throughout the host-rock.

Diamond pipes of South Africa, the Uranium minerals in the Singhbhum granites in Bihar (Jaduguda) are the examples.

(*b*) **Segregation.** This type of magmatic concentration is often due to the gravitative crystallization of early formed heavy minerals, *e.g.*, Bushveld chromite deposits (South Africa), chromite deposits of Keonjhar (Orissa).

(*c*) **Injection.** In this case, the metallic concentrates instead of remaining at the place of their original accumulation, get injected into the adjacent solid rock-masses It occurs at the residual magmatic stage, *e.g.*, Magnetite deposits of Kiruna (Sweden).

The late magmatic deposits are the consolidated parts of the igneous fractions that remained after the crystallization of the early formed rock-silicates. These deposits are formed by the following processes :

(*a*) **Residual liquid segregation.** Basic magmas undergoing differentiation may sometimes become enriched in iron and titanium. This residual liquid may drain out from the crystal interstices and consolidate without further movement. The host rocks are usually anorthosite, norite, gabbro etc, *e.g.*, Titaniferous magnetite bands of Bushveld complex.

(*b*) **Residual liquid injection.** In such cases, residual liquid may be squeezed out towards places of less pressure into the neighbouring rockmass or the interstitial liquid may be filter pressed out forming late magmatic injections, *e.g.*, Titaniferous magnetite deposits in Adirondack region of New York.

(c) **Immiscible liquid segregation.** Sometimes magma of an ore-and-silicate composition breaks down during cooling into two immiscible fractions which accumulate to form liquid segregation deposits, *e.g.*, Sulphide minerals usually associated with platinum, gold, silver copper etc.

(d) **Immiscible-liquid-injection.** The immiscible liquid accumulations before consolidation when subjected to disturbances, get injected into the surrounding rocks, forming immiscible liquid-injection. Nickeliferous sulphide deposits of Sudbury (U.S.A.) is an important example of this type.

2. Pegmatitic deposits. These are formed towards the very end of consolidation of the magma, in which the residual fraction is highly enriched with volatile constituents. Pegmatitic liquids may be squeezed out to fill in the cracks and fissures in the parent igneous body or the adjoining country rocks, and form pegmatite veins or dykes. These are usually formed between 500°C to 800°C.

Deposits of mica, feldspar, beryl, lithium minerals and tin mineral like cassiterite are included in the pegmatitic deposits. Besides, columbium, tantalum, thorium are the examples.

3. Contact metasomatic deposits. In the neighbourhood of invading magmas, alteration and replacement of the country rocks due to invasion of magmatic emanations may sometimes lead to the development of mineral deposits of economic importance. This process of formation of mineral deposits has been described as *pyrometasomatism* by Lindgren and as contact-metasomatism by Bateman.

In this case, the enclosing country rock is altered by the heat and other chemical constituents of the invading intrusive magma forming new minerals under conditions of high temperature and pressure. The deposits are usually resulted in calcareous rocks. The temperature of formation ranges from 400°C to 1000°C.

The gangue minerals in these deposits comprise an assemblage of high temperature metamorphic minerals, called '*skarn*', which are usually silicates of iron, magnesium, calcium and aluminium, depending upon the nature of country rock.

Thus deep seated batholithic masses of intermediate composition occurring within pure or impure carbonate country rocks serve as the most suitable locations, where the process of contact metasomatism can operate efficiently and lead to the development of mineral/deposits of economic importance. Thus contact metasomatic deposits are formed in nature. Examples are deposits of cassiterite, zinc, magnetite, graphite and sulphides of copper, iron, lead etc.

4. Hydrothermal deposits. As we know, a magma due to the alteration of physicochemical conditions gradually cools down, producir g rock-forming silicate minerals under different conditions of temperature and pressure. It also gives rise to a segregation of residual solutions enclosed within that parent rock, which is obvious from the Bowen's reaction series.

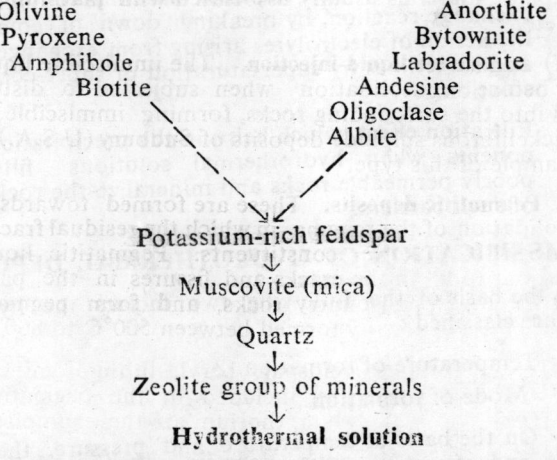

Olivine   Anorthite
Pyroxene   Bytownite
Amphibole   Labradorite
Biotite   Andesine
      Oligoclase
      Albite

Potassium-rich feldspar
↓
Muscovite (mica)
↓
Quartz
↓
Zeolite group of minerals
↓
Hydrothermal solution

Thus towards the end of the process of crystallization, the once widely dispersed gases and metals have collected near the top of the intrusive body. In moving up, the gases ooze and stream through the magma and collect some of the metals in their journey. At this stage pressure may force the gases and their dissolved rare elements to leave the magma chamber and to move along zones of weakness towards the surface. Such fluids may begin their journey upwards as liquids or gases which later becomes liquid and this hot water solution is known as hydrothermal solution.

Such hydrothermal solutions are important in the formation of certain kinds of mineral deposits, as they carry out metals from the consolidating intrusive to the site of deposition.

They are mostly epigenetic deposits. Since hydrothermal solution is originated from the magma, its temperature is about 350°C and it is under very high pressure. It is acidic in nature containing non-metallic and volatile constituents mostly. Metallic constituents of iron, zinc, nickel, gold, silver, lead etc. also find their places in the hydrothermal solution.

Causes of deposition :

1. Changes in the temperature.
2. Changes in the pressure of the system.
3. Exchange reactions between the substances in the solution.
4. Exchange reactions following mixing of solutions.

5. Exchange reaction between solution and wall rocks.

6. Changes in the pH of the medium (which determines the acidity/alkalinity of the medium).

7. Coagulation of the colloids, which is brough about by exchange reaction, by breaking down of complexion, by the action of electrolytes arising from exchange reactions, and sometimes by supersaturation or super-cooling of the solution.

8. Filtration effect, which helps in the precipitation of components when hydrothermal solutions filter through poorly permeable rocks and mineralize the rocks in front of such barriers.

CLASSIFICATION OF HYDROTHERMAL DEPOSITS

On the basis of the following two factors the hydrothermal deposits are classified :

(a) Temperature of formation.

(b) Mode of formation.

(a) On the basis of temperature and pressure, the depth of formation and the distance from the magmatic source, the hydrothermal deposits are of the following types :

(i) **Hypothermal deposits.** These deposits are formed at great depths, near the intrusive and within the temperature range of 300°C to 500 C

(ii) **Mesothermal deposits.** They are formed at a depth of 1500-4000 metres below the surface and within the temperature range of 200-300°C. The pressure ranges from 140 to 400 atmosphere.

(iii) **Epithermal deposits.** These are formed at shallow depths (further away from the surface). The temperature range is from 50°C to 200°C.

(iv) **Telethermal deposits.** These are formed under low temperature and pressure, far away from the parent igneous body with which their genetic relationship is not well established.

(v) **Xenothermal deposits.** These are formed by high-temperature ore forming fluids expelled from huge igneous rock masses, which have intruded into shallow depths. Thus they are characterised by high temperature, shallow depth of formation and rapid cooling.

(b) On the basis of the mode of formation, the hydrothermal deposits are of two types :

(*i*) Cavity-filling deposits.   (*ii*) Replacement deposits.

(*i*) **Cavity-filling deposits.** Cavities occur naturally in the earth as fractures or fractured zones along which the crust has been broken, sheared or displaced. In the cavities, the metallic minerals that are carried by the hydrothermal solution get deposited under epithermal condition.

Cavities are of two types :

1. *Original cavities*, which include pore spaces, crystal lattices, vesicles, lava drain channel, cooling cracks, bedding plane ard igneous breccia cavity.

2. *Induced cavities*, which include volcanic pipes, shear-zone cavities, solution caves, collapse breccia, tectonic breccia and cavities due to folding and warping.

Deposition of layers of different minerals upon the walls of a cavity leads to the development of crustification. Layers of crystals which are developed within the cavities, give rise to what is called *comb* structure. The open spaces left after the filling up of the cavities with mineral deposits are known as *vugs* or *drosses*. If cavity filling deposits are composed of one mineral only and devoid of crustification, they are said to be *massive*.

The cavity-filling deposits often include the fillings of fissure veins which range in shape, size and form between wide limits. The terms like stock-work, saddle-reef, ladder-vein, gash veins, pitches and flats etc. are used accordingly.

Masses of country-rocks are often enclosed within the fissure vein deposits and are known as *horses*.

(*ii*) **Metasomatic replacement deposits form,** when the hydrothermal solutions react with some mineral or substances in the crust, dissolving one substance and replacing it with the ore mineral. Mostly these deposits are formed under hypothermal condition.

The chemical composition and physical characteristics of the host rocks or minerals and the composition, temperature and pressure of the invading mineralising solutions determine the efficiency of the process as well as the extent and kind of replacement.

Metasomatic replacement deposits are characterised by :

(*a*) Presence of remnants of the country-rock.

(*b*) Presence of pseudomorphs of replacing minerals after the replaced ones.

(*c*) Absence of crustification etc.

5. Sublimation. This process is generally associated with volcanism and fumaroles. These are sublimates deposited at or near the surface at low temperature and pressure due to the sudden cooling of the vapours emanating from volcanoes or fumaroles. These deposits are generally very small and rather superficial and are seldom useful from the economic point of view. The sulphur deposits are the best examples of deposits formed by this process.

(B) Sedimentary processes of ore-formation. As we know, the combined action of sharp temperature fluctuations, wind, water freezing in rock crevices and plant roots penetrating into the rockmass, decomposes great blocks which gradually split into chunks, then into smaller fragments and are finally reduced to detritus and sand. In addition to mechanical destruction, primary minerals are subjected to chemical alteration, of which water is the principal agent. The products of weathering are transported by various geological agents, and finally the transported detritus gets deposited in various conditions of aqueous medium, to give rise to sedimentary formations. Five important processes are associated with the sedimentary mineralisation, *viz.*

1. Residual concentration.
2. Mechanical concentration.
3. Oxidation and Supergene Enrichment.
4. Remobilisation by meteoric circulating water.
5. Sedimentation.

1. Residual concentration. The term 'residual concentration' indicates the concentration of ores as residue. Residue left as *insitu* after weathering, followed by transportation sometimes give rise to valuable ore deposits. These are the insoluble products of rock weathering, the process which removes the undesired constituents of rocks or minerals. The residue may continue to accumulate until their purity and volume make them commercially important.

Of the three modes of weathering, the chemical mode of weathering is of paramount significance for the formation of residual deposits.

Factors Affecting the Residual-Concentration :

(*i*) **Climate**. The climatic conditions determine the course and intensity of rock-decomposition. Dittler has shown that a temperature above 20°C favours the process of chemical decomposition of bed rocks where SiO_2 goes into solution. The hot climate of the tropics creates the condition for residual concentration. Besides, an average amount of regular precipitation is most favourable for the formation of a mantle waste and subsequently residual concentration.

(*ii*) The presence of rocks or lodes containing valuable minerals of which the undesired constituents are soluble and the desired substances are generally insoluble under surface condition.

(*iii*) Long continued crustal stability is essential in order that residues may accumulate in quantity and that the residues may not be destroyed by erosion.

(*iv*) Existence of proper drainage is an important factor.

(*v*) There should be availability of reagents to bring about the breaking down of silicates and solution of silica.

(*vi*) **Local relief.** The optimum conditions for the formation of residual deposits are provided by the existence of an average hill-country topography, that ensures percolation of the meteoric precipitation down to the water table.

The relief must not be very great or the valuable residue will be washed away as rapidly as formed.

(*vii*) Since it takes much time to form mature, thick and well developed residual deposits, the acid rocks because of their relatively high alumina-content are the suitable rocks to produce residual deposits.

Under these conditions, for example, the feldspars of syenite or granite decomposes upon weathering to form 'bauxite', which persists at the surface, while other constituents are removed.

Hematite, manganese ore, bauxite, residual clays, ochres, kyanite deposits in India are some of the examples of residual concentration. Residual deposits are usually associated with laterite.

2. Mechanical concentration. It is the process of natural gravity separation of heavy from light minerals by means of moving water, or air by which the heavier minerals become concentrated into deposits called '*Placer-deposits*'.

As we know, during the process of weathering the resistant minerals like quartz, gold, platinum, magnetite, cassiterite, ilmenite, garnet etc. are set free in individual grains. Erosion now steps in and the detritus is swept down the slopes and into the water channels. Mechanical separation in running water or along sea or lake beaches sorts the detritus according to the specific gravity and the size of grains. The heaviest particles tend to collect in the lower part of the assorted detritus, the quartz grains are carried further, the minute and easily moved scales of clayey substances are ultimately deposited as sedimentary beds ; the colloids are coagulated by the electrolytes in the sea water.

Thus, two stages are there in the formation of placer deposits :

(*i*) The freeing by weathering of the stable minerals from their matrix, and

(*ii*) Their transportation and concentration.

Requirements for mechanical concentration :

(*a*) The valuable minerals should be of high specific gravity.

(*b*) They should be chemically resistant to weathering.

(*c*) They should be of adequate durability.

(*d*) There should be a continuous supply of placer minerals for concentration.

In the formation of placer deposits, nature employs in her own leisurely way the processes of crushing and concentration.

Factors Affecting Formation of Placer Deposits

(*i*) Specific gravity of minerals.

(*ii*) Specific surface of the particles.

(*iii*) Shape of the particles.

(*iv*) The ability of a body of flowing water to transport the particles and the viscosity of the transporting medium.

Types of placer deposits :

(*a*) **Eluvial placers.** Placer deposits along hill slopes are formed due to weathering and erosion of the country rocks containing low-grade deposits of the desired materials and are known as eluvial placers.

(*b*) **Deluvial placers.** When the weathered and disintegrated material is shifted down hills deluvial (scree or talus) placers are formed.

(*c*) **Proluvial (colluvial) placers.** Accumulation of the material at the foot of a slope can lead to the development of proluvial placers.

(*d*) **Alluvial placers.** Running water is the most important agency in the formation of alluvial placers. Irregularities on the floor of the channel in the form of natural barriers or riffles encourage deposition of placer deposits. Besides, at the meanderings of the river and at the confluence of tributaries, alluvial placers are formed.

(*e*) **Aeolian placers.** These are because of wind action, by which the lighter sand particles are blown away leaving behind a mass of coarser detritus containing valuable minerals.

(*f*) **Beach placers.** These are formed along the shores of lakes seas and oceans, mainly by the wave action.

There are essentially four factors which have much significance in the formation of placer deposits, they are :

(*i*) Geomorphological factors.

(*ii*) Climatic factors.

(*iii*) Hydrographic factors which is associated with the river action and the deposits at the meandering of the river and the junction between the tributaries.

(*iv*) Tectonic factors, which is associated with the rejuvenation of the base level of local and general erosion, creating conditions for recurrent cycles of erosional activity for development of alluvial placers.

3. Oxidation and supergene enrichment. This process is called as 'Infiltration' deposits also. This involves weathering and leaching of the upper parts of a mineral deposit (zone of oxidation) and redeposition of the ore-minerals at lower levels (zone of secondary or supergene enrichment).

The portion of the ore-body lying above the water-table is described as the zone of oxidation, since within this zone the ore-minerals forming the deposit may be oxidised readily in presence of air and water. By the reaction of the surface water containing free oxygen with the ore a solvent is formed. This solvent is very reactive and is helpful to oxidise the whole of the rock up to the water-table. Because of oxidation, there is a break-up in iron-sulphides (*e g.,* pyrite) giving rise to ferrous and ferric sulphates, sulphuric acid, sulphur and ferric hydroxide. Of these products ferric sulphate and sulphuric acid act as solvents and cause oxidation and solution of other ore minerals like galena, sphalerite, chalcopyrite, chalcocite etc.

Thus there are two main chemical changes within the zone of oxidation :

(*i*) The oxidation, solution and removal of the valuable minerals.

(*ii*) The transformation, *in-situ* of metallic minerals into oxidised compounds.

Since oxygen has no action on gold as well as other insoluble minerals like cassiterite, wolframite etc., these minerals remain in the outcrop without any change and are enriched upon contraction of volume. Much of the soluble minerals are naturally removed by running water. Newly formed compounds are precipitated due to reactions between solutions, by hydrolysis, by coagulations, as well as by reaction between the solutions and solids.

Factors controlling oxidation :

(*a*) **Water table.** Since oxidation takes place above the level of water table, the position of water table affects oxidation to a greater extent.

(b) Low-rate of erosion.

(c) Warm-humid climate with evenly distributed rainfall favours oxidation.

(d) To some extent the chemical composition of the rocks and the associated structural features also control oxidation.

Two types of deposits are mainly formed because of oxidation :

I. Above water table, there is oxidised deposits.

II. Deposits of supergene enrichment, formed below the water-table. In the case of the Ist type, the deposits are formed because of evaporation, saturation, and reaction between solutions bringing about their precipitation.

In the IInd type, the valuable minerals present in the oxidised zone, gets leached down because of oxidation. They begin to deposit in the secondary zone to make this zone enriched with ore, which is due to the fact that when the oxidised solution with minerals trickled down to the secondary zone which is below the water table ; there is no available oxygen and there they undergo deposition as secondary ores. The portion of the pre-existing ore-body, in which supergene sulphides are precipitated, is known as the zone of supergene-enrichment.

In the zone of supergene-enrichment the precipitation of the metallic sulphides is affected following Schurmann's law in the following order—silver, gold, copper, bismuth, lead, zinc, nickel, cobalt, iron etc. according to increasing solubility.

Gossan. Oxidation, solution and consequent downward movement of the valuable minerals lead to the concentration of useless residual materials and some of the dessicated products of oxidation upon the surface, where the ore-body had its outcrop and these together form a hard mantle known as *gossan* or *cap-rock*.

The gossan is made up principally of limonite, gangue minerals and some of the oxidised products of the ore minerals. Sometimes, false gossans are, however, produced as a result of precipitation of extraneous ferruginous solutions upon the exposed surfaces of the country-rocks.

But in the majority of cases, gossans supply many decipherable inferences as to the size, character and mineral contents, of the hidden ore deposits. Therefore gossans are considered as sign boards of oxidised as well as enriched zones beneath the surface.

Thus the process of oxidation and secondary enrichment produces the following :

(*a*) Gossan.

(*b*) Zone of oxidation and leaching.

(*c*) Secondary zone of enrichment.

(*d*) Zone of unriched primary ore etc.

4. **Process of remobilisation by circulating meteoric water.** This process involves the solution of materials dispersed in the adjacent rocks by ground water and their subsequent concentration under suitable physico-chemical conditions.

This process operates at moderate depth to shallow depths, under moderate pressure and low temperature (ranging between 0°C to 100°C). These deposits are formed by ground water, and occur mainly in sedimentary rocks like—limestones, sandstones and shales, in the presence of some reducing agents.

The occurrence of some manganese deposits in the Keonjhar District of Orissa are believed to have formed by this process where by manganese at one level has been remobilised to the other and its subsequent concentration at that level. Other examples are—phosphates, barite, celestite, pyrite, flint etc.

5. **Sedimentation.** The process of sedimentation gives rise to following types of deposits :

(*i*) **Evaporation deposits.** These are formed by evaporation of surface waters giving rise to deposits of salt, anhydrite, gypsum, borates, nitrates etc.

(*ii*) Chemically deposited materials, which is due to reactions between the solutions in bodies of surface water, are sometimes resulted in the process of sedimentation. Calcareous, ferruginous, manganiferous deposits are usually due to such processes.

(*iii*) **Organic deposits.** Accumulation of organic matter whether due to chemical reactions or due to as such transportation and deposition by the geologic-agents give rise to deposits like bog iron ore, coal, phosphatic, siliceous as well as calcareous deposits of chalk, fossiliferous limestones etc.

The mechanically deposited materials like placer deposits are also due to sedimentary processes. The diamondiferous-conglomerates as occur in the Vindhyan system of India are the examples of such deposits.

Thus 'sedimentation' gives rise to deposits both through mechanical means as well as through chemical reactions and also through organic—chemical actions (*i.e.*, by bio-mechanical + biochemical actions).

(C) Metamorphic-processes. As we know, three important factors like temperature, pressure and chemically active fluid play dominant roles in bringing about transformation of the original rock and mineral deposits. These metamorphic changes are sometimes responsible in giving rise to important mineral deposits, which are conveniently grouped into two classes :

(*i*) **Metamorphosed deposits.** These are formed as a result of metamorphism of pre-existing mineral-deposits, due to which the ore acquire new-textural characteristics like granular, schistose, banded etc.

Through the process of metamorphism the original minerals usually get anhydrated, and transformed into new-minerals. Thus limonite and other iron-hydroxides are transformed into haematite and magnetite ; psilomelene and manganite are replaced by braunite and hausmannite etc. Certain ferruginous and manganiferous sediments undergo metamorphism and give rise to iron and manganese ore deposits.

It may also so happen that during the process of metamorphism, the originally dispersed ore mineral get segregated giving rise to workable-economic deposits.

(*ii*) **Metamorphic mineral deposits.** These deposits are formed due to recrystallisation and/or recombination of certain mineral constituents of the rocks during regional or contact metamorphism. Regional progressive metamorphism has the greatest significance in the formation of metamorphic deposits. The shape, constitution and composition of mineral bodies are altered in consequence of it.

The most common examples of the non-metallic minerals which are formed through this process are graphite, garnet, kyanite, sillimanite, asbestos, steatite, etc.

Metallic minerals like corundum, rutile, ilmenite, magnetite, uranium, pyrite, native copper (Jake superior type) etc.

The type of mineral deposits resulted are dependent on the facies of metamorphism.

PART VII

ORIGIN, MODE OF OCCURRENCE AND DISTRIBUTION (In India) AND ECONOMIC USES

CHAPTER 66

GOLD

The economic gold minerals are :

(i) Native gold (metallic gold).

(ii) Gold amalgam (gold alloyed with silver).

(iii) **Tellurides.** Sylvanite, calaverite, petzite and nagyagite.

Origin. All primary gold deposits are formed usually during the last stages of the chilling of the magma, which rise along fissures to the upper layers of the earth's crust from great depths. The gold is transported together with the magma from the depths of the earth in hot aqueous solution and vapours. In due course, these solutions get solidified with the falling of temperature, giving rise to ore bodies mostly in the form of veins filling fissures in rocks.

These veins generally consist of quartz with a small admixture of other minerals and particles of gold in the form of fine grains, platelets and flakes and sometimes crystals, wires, filaments and so on. Big chunks of gold are called 'nuggets'. The tellurides and auriferous sulphides also occur in veins.

In the course of time, these rocks and veins under the action of various geological agents undergo mechanical disintegration and chemical decomposition. The weathered products while being carried down to the depositional site undergo the processes of mechanical concentration and all these processes give rise to *placer-gold* deposits. The factors contributing to the concentration of gold in placers are its high specific gravity and resistance to chemical alteration through weathering.

Mode of occurrence. Most of the primary gold deposits occur in the vicinity of acidic igneous intrusions and have been formed by hydrothermal solutions as *replacement* or *cavity-filling* and a few have been formed by *contact metasomatism*.

In India, gold occurs as primary veins or lodes or as *placers*.

The other modes of occurrence of gold deposits in various parts of the world are as follows :

(*i*) **Magmatic deposits.**   South Africa (Waarkral).

(*ii*) **Contact metasomatic.**   Utah (U.S.).

(*iii*) **Residual concentration.**   Brazil, Australia.

Distribution in India :

1. **Kolar-gold-field.** It is situated in the state of Karnataka and has four productive mines—Nandydroog, Champion Reef, Mysore, and Ooregaum. The Champion Reef mine is the deepest mine in the world. In these cases, the lodes are localised along or in vicinity of stratigraphic contacts of granular, fibrous and amphibolites of the schist belt. The lodes are made up of quartz-vein zones in altered and mineralised wall rock. The mineralising solutions responsible for the development of the auriferous veins of South India, were probably derived from the magma which gave rise to champion gneisses. The principal process, which led to the development of the auriferous veins, was cavity-filling and metasomatic replacement of the wall-rocks is said to have assumed a less significant role.

2. **Hutti gold-field.** In Hyderabad, auriferous veins occur in the Hutti-gold-field within the schistose rocks of Dharwarian age.

3.   Ramgiri gold-field (Andhra Pradesh).

4.   Wynaad gold-field (Partly Tamilnadu and Partly Kerala).

5.   Chotanagpur region, Kundra-Kocha Gold-field.

The alluvial placer gold deposits occur in Assam (Subansiri river), Bihar (Subernarekha), the Son, Deo and Ib-rivers of Madhya Pradesh, in Koraput and Sambalpur districts of Orissa etc.

Economic uses :

(*i*) Gold is a precious metal and constitutes the international standard of exchange.

(*ii*) In alloy-state with other metals, it is mostly used in ornamentation.

(*iii*) The industrial uses of gold alloys are in dentistry, chemical plants, thermocouples, watches, X-ray equipment, photography and in some medicines.

CHAPTER 67

IRON

Being one of the most widely distributed elements in the earth's crust iron rarely occurs in the free state as it enters into the composition of many rocks and minerals. It consists of about 4·6 percent of the earth's crust. In nature, iron occurs in four principal forms, *viz.*, oxides (hydroxides also), carbonates, sulphides and silicates. The chief economic iron ore minerals are

Magnetite	Fe_3O_4	(Containing 72·4% of iron)
Hematite	Fe_2O_3	(Fe=70%)
Limonite	$2Fe_2O_3, 3H_2O$	(Fe=59·8%)
Goethite	Fe_2O_3, H_2O	(Fe=62·9%)
(Spathic-ore)- Siderite	$FeCO_3$	(Fe=48·2%)
Pyrite	FeS_2	(Fe=46·2%)

Chamosite and thuringite are examples of iron-silicate minerals.

Origin. Iron ore deposits are generally formed by the following processes :

(*i*) **Magmatic.** Magnetite, titaniferous magnetite, *e.g.*,—Kiruna (Sweden), Keonjhar and Mayurbhanj (Orissa), Salem (Tamilnadu), Hassan (Karnataka) etc.

(*ii*) **Sedimentary.** Hematite deposits of Bihar, Orissa, Madhya Pradesh, Maharashtra and Karnataka. Siderite deposits of economic importance are usually sedimentary deposits.

(*iii*) **Replacement.** Magnetite, hematite, deposits, *e.g.*, Lyon Mountain, Newyork.

(*iv*) *Residual concentration.* Laterite formations in the Eastern ghats of India.

(*v*) *Oxidation.* Limonites, *e.g.*, Rio Tinto, Spain.

Besides the above, contact metasomatism also plays some role in the formation of magnetites and specularite.

Regarding the origin of vast deposits of Banded-Hematite-Quartzite and Banded Hematite-Jasper, there is a divergence of opinion and several theories in this connection have been propounded ; the most important ones are as follows :

(*a*) **Chemical precipitation theory.** Workers like Jones, Percival and Krishnan suggest chemical precipitation to account for the origin. According to them, iron-oxide and silica were deposited under submarine conditions by rhythmic chemical precipitation from meteoric as well as magmatic water.

(*b*) Wazalwar and Nandy regard the ferruginous shales to be responsible in the formation of iron-ore deposits by processes of *leaching, replacement and concentration* of iron oxides and hydro-oxides by action of circulating meteoric water, aided in certain cases by igneous action.

(*c*) Dunn believes that BHQ and BHJ are formed by the secondary silicification by thermal activity of the volcanic tuffs and flows now represented by ferruginous, chloritic or carbonaceous shales or phyllites of the area.

(*d*) Sakamoto attributed the cause of banding in the precambrian iron ores to a cyclic deposition of colloids due to a *periodic change of pH value* in the depositional medium. He points out that iron migrates in an acid environment and precipitated in a neutral or passive environment, whereas SiO_2 migrates in an alkaline medium and is precipitated in acid environment.

Mode of occurrence. Iron-ore deposits occur as magmatic deposits, as bedded-deposits, as residual concentration deposits or sometimes as nodules and concretions in shales associated with coal-seams.

Distribution in India :

(*i*) The biggest iron-ore field of India is situated in the Singhbhum district of Bihar and the adjoining districts of Keonjhar, Sundergarh and Mayurbhanj of Orissa. The *massive-ores* where the iron content ranges from 66 to 70% occur on top of hill ranges. The *shaly* ore may be as rich as hematite and as low as 50% or less in iron. The *biscuit* or *laminated* ore contain about 55 to 66% of iron. *Blue dust* ore, which is an extremely friable and micaceous hematite powder contain about 68% of iron, is formed by leaching process from BHQ. There also occurs lateritic ore.

The important mining centres of Orissa and Bihar are Barbil, Gua, Bonai, Joda, Kiriburu, Suleipat, Gorumahisani, Noamundi, Barajamda etc.

(*ii*) **Madhya Pradesh.** In the Bailadila hill ranges.

(*iii*) **Maharashtra.** Ratnagiri district.

(*iv*) **Goa.** **Bicholim**—Pale in Goa.

(*v*) **Karnataka.** Bababudan hills in Chikmagalur district, and in Sandur, Bellary, Hospet districts as well as Shimoga and Chitaldrug districts. Important one is that of 'Kudermukh'.

(*vi*) **Andhra Pradesh.** Cuddapah, Kurnool, Chitoor, Nellore, Anantapur, Warangal and Adilabad districts.

(*vi*) **Tamilnadu.** Salem district, and Tiruchirapalli district.

(*viii*) **West Bengal.** Deposit of lateritic ores mostly occur in West Bengal.

(*ix*) **Assam.** Iron stone-clay are found as nodules and thin beds in the coal measures of Eocene age and in the Tipam saries of Miocene age.

CHAPTER 68

COPPER

It is the most important non-ferrous metal and was the earliest metal used by man.

Ore-minerals. In nature copper occurs in four principal formes, *viz.*, sulphides, carbonates, oxides and as native copper. Of these the bulk of copper is obtained from the sulphide ores. The chief economic ore-minerals however, are

		Composition	*% of copper*
1.	*Native copper*	Cu	100
2.	*Sulphides* :		
	(i) Chalcopyrite	$CuFeS_2$	35·5
	(ii) Bornite or erubescite	Cu_5FeS_4	63·3
	(iii) Covellite	CuS	66·4
	(iv) Chalcocite	Cu_1S	79·8
	(v) Enargite	Cu_3AsS_4	48·3
	(vi) Tetrahedrite	$Cu_8Sb_2S_7$	52·1
3.	*Carbonates* :		
	(i) Azurite	$2CuCO_3Cu(OH)_3$	55·1
	(ii) Malachite	$CuCO_3Cu(OH)_2$	57·3
4.	*Oxides* :		
	(i) Cuprite	Cu_2O	88·8
	(ii) Tenorite	CuO	79·8
	(iii) Chrysocolla	$CuSiO_3H_2O$	36

To be economically exploited a copper ore should contain at least 2·5% of copper. In modern times ores with 1% of copper are also used.

Origin. All large copper ore bodies are closely connected with igneous rocks mostly of an acidic nature. It is mostly believed that copper deposits are formed through hydrothermal solutions, either as cavity-filling or replacement deposits. But replacement has been a more dominant process than cavity filling.

Only a few deposits have been formed by magmatic concentration or by contact-metasomatism.

The process of oxidation and supergene enrichment also plays dominant role in giving rise to workable deposits of secondary copper sulphide ores :

Mode of occurrence. Copper deposits may occur as

(*a*) **Disseminated ore bodies.** Where the copper minerals are generally dispersed in a large volume of rock. They are generally of low grade. The porphyry-copper deposits of USA are of this type.

(*b*) **Massive, irregular or lenticular** ore bodies, which are formed by the process of replacement.

(*c*) **Vein deposits or lodes.** In which the copper bearing solutions percolating along shear-zones and rock-fractures deposit copper minerals with changes of temperature and pressure forming fissure-veins, *e g.,*— Copper deposits of Singhbhum.

(*d*) Deposits following stratigraphic beds, as is the case with the deposits of Khetri (Rajasthan).

Distribution :

(*i*) In Andhra Pradesh, the most important copper deposits are the Agnigundla-deposits.

(*ii*) In Bihar, in the Singhbhum district, a copper bearing belt of about 80 miles long occurs. Here the copper ores occur as veins in the country rock consisting of mica-schists, quartz-schists, chlorite-schists, biotite-schists, granite and granite-gneisses. The veins are best developed along a zone of over thrust, where they form well defined lodes, as seen at Rakha mines, Mosabani and Dhobani. Individual lodes normally consist of one or more veins one inch to two feet thick, the average being 5 to 7 inches.

(*iii*) In Madhya Pradesh, the important deposit is the Malan Jhakhand copper deposit, where copper ores occurs in the form of veins within dolomitic limestone.

(*iv*) **The Khetri copper deposit of Rajasthan** is one of the important copper deposit in the country. This belt has 3-richly mineralised sections—Madhian, Kolihan and Akhwali. The copper ore bodies occur in phyllites, slates and schists of the Ajabgarh series (Delhi system) as irregular stringers, fillings of schistose planes and fractures and disseminations in the host rock. The mineralisation in Rajasthan copper belt is epigenetic and seems to have occurred under mesothermal conditions from post-Delhi (Erinpura) granite magma.

(*v*) Other important copper deposits of the country are as follows :

(*a*) *Himachal Pradesh.* Kangra, Kulu valley.

(*b*) *Mysore.* Chittaldrug, Hassan, Bellary districts.

(*c*) *West Bengal.* Darjeeling, Jalpaiguri districts.

(*d*) *Sikkim.* Rangpo and Dickchu deposits which are found to occur in association with the metamorphic rocks belonging to the Daling series.

Economic uses. The metal is of great industrial importance, because of its high electric conductivity, high ductility and malleability. Thus it is mostly used in electrical manufactures. Besides, the copper alloys are used in buildings, automobiles, air planes, naval ships, house hold utensils as well as in metallurgy and paints.

CHAPTER 69

MANGANESE

The chief sources of manganese are the oxide-minerals. There are a total of 156 manganese minerals of which 44 are found in India. According to the composition of manganiferous ores in regard to the proportion of manganese to iron, it is customary to use the term 'manganese ore' to those containing over 40% of manganese. The most common economic minerals are

1. Pyrolusite MnO_2 Mn 63·2%

    (the harder coarsely crystalline variety is termed polianite which occurs in groups of needle-like crystals)

2. Hausmanite Mn_3O_4 72·5%
3. Braunite Mn_2O_3 64·3%
4. Manganite Mn_2O_3, H_2O 62·4%
5. Psilomelane $Mn_2O_3, 2H_2O$ 45 to 60%

    (the earthy, loose and soft variety of impure psilomelene with pyrolusite is called-wad)

6. Rhodochrosite $MnCO_3$ 47·6%
7. Rhodonite $MnSiO_3$ 41·9%.

Other important manganese minerals which are of specially Indian origin are as follows :

Sitaparite, Hollandite, Jacobsite, spessartite etc.

Origin. Manganese ore deposits may be formed as follows :

1. Hydrothermal deposits. It is formed by magmatic hot-water solution.

2. Sedimentary deposits. Due to chemical precipitation.

3. Residual deposits. Due to residual concentration.

4. Metasomatic replacement. Through the action of underground water containing manganese.

5. Metamorphosed deposits. Due to metamorphism of the above said deposits.

On the basis of their mode of occurrence and association with different kinds of country-rocks, however, the Indian manganese ore deposits have been classified as :

(*a*) **Gonditic ores.** Which are associated with metamorphosed manganiferous sediments.

(*b*) **Koduritic ore.** These are produced due to reaction between the country-rocks and an invading magma of granitic composition. The hybrid rocks, thus produced are called *kodurites*.

(*c*) **Lateritoid ores.** These are produced due to metasomatic replacement and residual concentration.

Mode of occurrence. Manganese deposits occur as bedded sedimentary deposits, metamorphosed deposits, residual deposits or hydrothermal deposits.

Distribution in India. In India, extensive and rich manganese deposits occur in Madhya Pradesh, Orissa, Bihar, Andhra Pradesh, Maharashtra and Karnataka.

According to the genetic classification of 'Indian Manganese Deposits' by Supriya Roy the followings are the types of deposits and their area of distribution :

1. Syngenetic deposit :

(*a*) Regionally metamorphosed manganese sediments, associated with pelitic and psammitic rocks and very often with manganese-silicate rocks (Gondites). Example : Mansar formation of Sausar group of M.P. and Maharashtra.

(*b*) Regionally metamorphosed manganese sediments associated with marbles and calc-silicates, *e.g.*, Mn-ore bodies in Lohangi marbles of Sausar group of MP and in calc-silicate rocks of khondalite group in Andhra Pradesh and adjacent parts of Orissa (Kodurites).

(*c*) **Contact-metamorphic deposits of Gujarat.** Due to thermal action of granitic intrusive on pre-existing manganese sediments, associated with impure limestone.

2. Epigentic.

(*a*) Redsidually concentrated deposits of Goa, Keonjhar-Bonai (Orissa) etc.

(*b*) **Replacement and cavity-filling deposits.** By meteoritic water, through precipitation from 'gels' or solution.

Other important distributions are :

(*i*) *Bihar*. Barjamda and Singhbhum district.

(*ii*) *Karnataka*. Residual deposits occurring within the country rocks of Dharwarian age.

Economic uses :

(*i*) It is one of the important ferro-alloy metals.

(*ii*) Wide application in steel industries.

(*iii*) In chemical industries for dry batteries.

(*iv*) As a decolouriser in glass industry and also as oxidising agents.

The manganese ores, on the basis of their manganese contents may be classified into the following grades.

(*a*)	Chemical grade	82-87% of Mn.
(*b*)	Metallurgical grade :	
	First grade	more than 48% of Mn.
	Second grade	45-48% of Mn.
	Third grade	less than 45% of Mn.
(*c*)	Manganese ore grade	35 to 45% of Mn.
(*d*)	Ferruginous Manganese ore grade	10 to 35% of Mn.
(*e*)	Manganiferous iron ore grade	less than 10% of Mn.

For trade purposes, Indian Mangnese ores are classified as :

(*a*)	Battery grade	80-86% of MnO_2.
(*b*)	Peroxide grade	78% MnO_2+4% Fe.
(*c*)	High grade	46 to 48% of Mn.
(*d*)	Low grade	38 to 40% of Mn.
(*e*)	Ferruginous grade	30-35% of Mn.

In the paints and pigments as well as in Fertiliser industries. also manganese is used.

CHAPTER 70

CHROMIUM

It is an important alloying element in the manufacture of steel. Chromite is the only ore-mineral of chromium.

Chromite—FeO, Cr_2O_3, $Cr_2O_3 = 68.0\%$ and $Cr = 46.66\%$.

Origin. Chromite deposits are magmatic segregations in ultrabasic igneous rocks of Archaean age. Chromite is associated only with highly basic or ultrabasic rocks like peridotite, saxonite, dunite and pyroxenites or their alteration product, serpentine rock.

Mode of Occurrence. Chromite deposits occur as lenses, masses, veins and disseminated grains in host rocks. The deposits are regarded as the early or late magmatic segregation or injection product.

Distribution in India. The largest chromite deposit in the country is located in the Sukinda-ultrabasic belt of Cuttack and Dhenkanal districts of Orissa, and also in the Keonjhar district of the state. The belt extends over a distance of about 20 km. the width of the belt is about 2 km. The ore bodies are lenticular in shape, and occur as lenses and patches within the lateritised ultrabasic rocks. The following type of ores are found to occur viz.

(i) massive ore   (ii) banded ore
(iii) disseminated ore   (iv) ferruginous lateritic ore
(v) powdery or friable ore   (vi) conglomeratic ore
(vii) placer ore etc.

The reserve is estimated to be about 8 million tonnes.

The other important deposits occur in :

(i) **Andhra Pradesh.** Kistna district (Kondapalle).
(ii) **Bihar.** Singhbhum district.
(iii) **Karnataka.** Chitaldrug, Hassan and Shimoga districts.
(iv) **Tamil Nadu.** Salem districts (Sittampundi).

Economic uses :

(i) In the metallurgical industries in the production of various non-ferrous alloys of chromium and also in the form of ferro-chrome for manufacturing chrome steel.

(ii) In refractory industries, due to its high resistance against corrosion, high temperature and sudden temperature changes and its chemically neutral character.

(iii) In chemical industries, for the manufacture of chromium compounds like chromates and bi-chromates and chromic acid etc.

CHAPTER 71

ALUMINIUM

Aluminium constitutes about 8·07 percent of the earth's crust and is the most abundant element next in importance to oxygen and silicon. The only ore, from which aluminium is extracted is called '*bauxite*', which is not a mineral but an aggregate chiefly of gibbsite, boehmite, and a little of kaolinite.

'Bauxite' is a secondary product in between monohydrate diaspore and tri-hydrate gibbsite, *i.e.*, between $(Al_2O_3H_2O)$ and $(Al_2O_3, 3H_2O)$. Thus it is a generic term for rock rich in hydrous aluminium oxide $(Al_2O_32H_2O)$. The ore bauxite has a typical oolitic or psiolitic structure and it is also found in an amorphous form. The common aluminous minerals which are found to occur in association with bauxite are Gibbsite (hydrogillite) Diaspore, and Bohemite.

Presence of about 50% of Al_2O_3 is considered to be the lower limit of a good commercial bauxite.

Origin :

1. Bauxite deposits are sometimes considered to be the outcome of chemical sedimentation. According to some geologists with continued weathering, the sodium, calcium, potassium, magnesium and iron content of magmatic rocks separated leaving behind aluminium and silica to form kaolinite.

Sulphuric acid, the usual product of the oxidation of pyrite and other sulphides, destroys the strong bond between aluminium and silica of aluminosilicates and aluminium is separated in form of soluble salts. Dilution by atmospheric water lessen the acidity of sulphuric acid solution, thus making possible the precipitation and depositions of aluminium of solution. Then, again the organic acid present in the water of most rivers and lakes combine with aluminium to form soluble compounds which are transported by rivers over large distances without being precipitated.

Because of the various geochemical mobility of iron, manganese and aluminium compounds, they become differentiated in the 'inshore-zone' of basin. Solubility studies have shown that pH 7 to 9·5 is conducive to the formation of bauxite. The formation of bauxite occurs in sea water, the pH value of which varies in relation to depth. Bauxite begins to accumulate nearer to the shore.

2. Residual concentration It is the process whereby ore deposits are formed as residue, left as *in-situ* after weathering followed by transportation. The residues are the insoluble products of rock weathering which have escaped distribution by transporting agencies and which still mantle the rock from which they have been derived.

As we know, under ordinary conditions of weathering and humid temperate climates, the silicates of the alkalies, lime and aluminium, (*e.g.*, feldspar) lose their base and gain water, forming hydrous aluminous silicates.

As Cooper states "it is probable that water, carbon-dioxide, humic acid and tapid rainwater are the reagents involved". Carbonic acid and organic acids break up silicates and the resulting alkali carbonates are competent solvents. The chemical reactions which usually take place are as follows :

$$(i) \ 6H_2O+CO_2+2KAlSi_3O_8 \rightarrow \begin{cases} Al_2Si_2O_5(OH)_4=\text{clay} \\ + \\ 4SiO(OH)_2=\text{silicic acid} \\ + \\ K_2CO_3=\text{alkali} \end{cases}$$

$$(ii) \ 2 \ KAlSi_3O_8+H_2CO_3+H_2O \rightarrow K_2CO_3+Al_2Si_2O_5(OH)_4 \\ +2SiO_2 \text{ (in-solution)}$$

$$(iii) \ nH_2O+Al_2Si_2O_5(OH)_4 \rightarrow Al_2O_3, \ nH_2O+2SiO(OH)_2 \\ \text{(bauxite)}$$

Under conditions of poor drainage clay minerals like kaolinite, illite, montomorillonite etc. are formed. If there are well defined wet and dry season and fairly good drainage, the clay minerals are decomposed to form laterite. But with an evenly distributed rainfall, all through the year, ironoxide tends to be separated in solution, because of aeration, leaving behind mainly aluminium hydroxide which is called 'Bauxite'. When the aluminous constituents predominates the colour lightens to yellowish or whitish and the rock becomes early or clay-like.

Mode of occurrence. Bauxite deposits occur as

(*a*) **Blanket deposits.** That occur at or near the capping cr as flat or undulating sheets or lenses under some soil cover.

(*b*) **Interstratified deposits.** Which lie on erosional surfaces and invariably occupy unconformities.

(*c*) **Pocket deposits** . These deposits are restricted to limestone or dolomite. They are shaped like huge teeth with many projecting roots.

(*d*) **Detrital deposits.** Which are produced from the pre-existing deposits. They may be talus accumulations, stream gravels or more or less unconsolidated low level surface layers.

Distribution in India. The richest deposits of bauxite are commonly associated with laterite, which occur as blankets or cappings in the high plateaux of our peninsula. Plateaux at an altitude of 900 to 1000 metres are generally regarded as good homes of bauxite deposits.

Andhra Pradesh. Vishakapatnam, East and west Godavari district.

Bihar. Ranchi and Palaman districts (Lohardaga).

Madhya Pradesh. Katni and Amarkantak plateau.

Tamil Nadu. Shevaroy hills bauxite of Salem.

Maharashtra. Kolba bauxite.

Gujarat. Jamnagar and Kaira deposits.

Orissa. Extensive and huge deposits of bauxite are found to occur in Koraput, Kalahandi, Bolangir and Sambalpur districts.

Economic uses :

(*i*) In the manufacture of alum, aluminous sulphates and other chemicals.

(*ii*) In the construction of air-planes, automobiles, electrical appliances etc.

(*iii*) In the manufacture of containers, utensils and machineries etc.

CHAPTER 72

LEAD AND ZINC

The two metals lead and zinc rarely occurs in native state, they generally occur in combination with other elements. The ore minerals of lead and zinc are usually found to occur in association with each other. The followings are the important minerals of lead and zinc :

Lead	Zinc	
Galena-PbS-Pb 86·6%	Sphalerite Or Zinc blende	ZnS, Zn-67%
Cerussite-PbCO$_3$-Pb 77·5%	Smithsonite Or Eng. Calamine	ZnCO$_3$, Zn-52%
Anglesite-PbSO$_4$, Pb 68·3%	Hemimorphite Or Americanname- Calamine Zincite-ZnO	2ZnO$_2$SiO$_2$2H$_2$O, Zn-54·2%

Origin. Lead and zinc ore minerals, particularly the sulphide-ones are formed due to contact metasomatism, replacement by hydrothermal solutions.

Mode of occurrence. Most of the lead ore mines of the world are also zinc ore producers and nearly all zinc ore deposits carry lead ore. Both lead and zinc ore bodies usually occur as veins and massive or tabular lodes, and as disseminations, mostly in limestone or dolomites. Majority of these ores occur as cavity-fillings and replacements formed by low-temperature hydrothermal solutions.

Distribution in India. The most important lead-zinc deposits of economic value in India is the Zawar deposit of Udaipur district of Rajasthan. India's reserve of these ores is meagre compared to her needs.

In the zawar area, the Mochia Magra, Barai Magra and Zawar Mala hills contain most extensive deposits.

The Zawar mine is located in the Mochia Magra hills. The principal rock types of zawar area consists of phyllites, slates, mica-schists, dolomites and quartzites of the Aravalli system. But mineralisation of lead and zinc sulphides is solely confined to the dolomites, whereas adjoining phyllites are almost barren. The lead and zinc deposits are confined to the upper series of the Aravalli-rocks in the zawar area. The localisation of the ores are structurally controlled by the shear zones. The ores occur in shear zones and follow shear planes which are the youngest tectonic feature in the area.

Most of the ore-shoots are found to occur as irregular steeply dipping and thin parallel tabular masses. Galena is generally concentrated in some particular portions of the deposits but the sphalerite is more or less evenly distributed. The gangue is dolomite and quartz.

Evidences obtained so far suggest the formation of the Zawar lead-zinc deposits from hydrothermal solutions at about 250°C.

The ore minerals consist of argentiferous galena associated with sphalerite containing a small percentage of cadmium, pyrite, arsenopyrite and chalcopyrite. The ore contains 1·5 to 2% of lead and 4·5 to 5% zinc.

Other important occurrences in the country are as follows :

(a) Lead copper ore deposits in Agnigundla area of Guntur district of Andhra Pradesh.

(b) Lead-zinc copper belt of 3-km long in Ambamata-Devi area of Gujarat and Rajasthan.

(c) Sargipalli area in the district of Sundergarh (Orissa).

The estimated reserve in the country is about 9 million tonnes.

Economic uses :

(i) Lead is used in the construction of accumulators, for lead piping and sheeting, cable covers, as pigments in glass making, in medicine etc.

(ii) Zinc is used for coating, galvanising iron and steel products, in the manufacture of pigments and alloys with other metals (like brass, bronze, german silver), in the manufacture of batteries and electric appliances. Besides, they are widely used in textile industry, timber preservation etc.

CHAPTER 73

MICA

The term 'Mica' covers a large group of rock-forming minerals. Natural mica forms hexagonal crystals of varying size, the distinguishing feature of which is their ability to split readily into separate leaves.

Minerals. Minerals of the mica-group are composed of the orthosilicates of aluminium, with potassium and hydrogen and generally magnesium, iron, sodium and rarely rubidium and caesium. The main mica minerals are the following :

(i) Muscovite, also known as white-mica or potash-mica.

(ii) Biotite, known as black-mica or magnesium-iron mica.

(iii) Phlogopite, *i.e.*, Amber mica or magnesium-mica.

(iv) Lepidolite—also known as lithium-mica.

(v) Zinnwaldite, also called lithium-iron mica.

(vi) Roscoelite, (vanadium-mica).

(vii) Fuschsite, *i.e.*, chrome-mica.

Origin. The micas, which are usually associated with acid igneous rocks, are formed towards the end of the process of crystallization of rock-forming silicates. They are mostly formed as pegmatite deposits, consisting of very coarse grained igneous rocks occurring as dykes or veins.

Mode of occurrence. Mica-minerals occur in igneous, sedimentary and metamorphic rocks formed under different geological conditions. While muscovite occurs in pegmatites of acidic nature, phlogopite mica is restricted to basic-pegmatite. Lepidolite occurs in pegmatites associated with topaz. Commercial biotite is found to occur mostly in biotite-schists.

Distribution. India is the most important mica-producing country in the world and it supplies about 80% of the world requirements of block-mica. The occurrence of muscovite-mica is associated with the rocks of Archaean age. The three most important cocurrences are (*i*) the Koderma Mica Belt in Bihar,

(*ii*) the Nellore Mica Belt in Andhra Pradesh,

(*iii*) the Rajasthan Mica Belt in Rajasthan.

Koderma-mica-belt. It is about 32 km wide and stretches from Gaya district through Hazaribagh and Monghyr to Bhagalpur district for about 145 km. In this mica belt, the deposits of mica are associated with the pegmatite veins which traverse through the schistose and gneissose country rocks. The pegmatite veins contain mica-deposits at places where they traverse through the mica-schists. The blocks of muscovite which occur within the Bihar-mica-belt are generally reddish in colour and are, therefore, known as 'Ruby-mica'.

(*ii*) **Nellore-mica-belt.** It has a length of about 100 km. between Gudur and Sangam. The country rocks are Archaean-mica-schist and Hornblende-Schist which are intruded by pegmatite veins. Here muscovites are light green in colour.

(*iii*) **Rajasthan-mica-belt.** It is a quite wide belt and produces 19% of the total Indian-production. Here the mica bearing peg-matites are intrusive mainly to rocks of the gneissic complex and also into Aravalli-schists.

Uses :

(*i*) As an insulating material in electrical industry.

(*ii*) Muscovite, phlogopite splittings are used in making of build-up-mica or micanite and other insulation product, for both heat insulation and also electrical insulation.

(*iii*) In powder-form it is used in lubricating oils and decorative wall papers.

CHAPTER 74

GYPSUM

Gypsum is a hydrated calcium sulphate, which crystallizes in the monoclinic system. The mineral is having the chemical composition as $CaSO_4.2H_2O$, where $CaSO_4$ constitutes about 79.1% of the mineral and water is about 20.9%. The material should not be called as gypsum, if it contains less than 64.5% of $CaSO_4$, $2H_2O$ by weight. It belongs to the class of mineral deposit called 'Evaporites'.

Varieties. There are five varieties of gypsum as

(a) Pure gypsum or salenite, which is crystalline in character and is transparent.

(b) Albaster, a dense, massive, granular, transluscent variety.

(c) Satin-spar, a fibrous variety having silky lusture.

(d) Gypsite, an earthy, soft, impure variety containing abundant small gypsum crystals scattered through clayey or sandy-soil.

(e) Rock-gypsum, a coarse granular, compact, massive variety, which usually occurs interbedded with sedimentary rocks, and is usually impure.

Origin. It is believed that gypsum is formed by the action of sulphuric acid produced by the oxidation of pyritic matter, on limestone and marl. For this reason, well developed crystals and plates of gypsum are found associated with clay and limestone. It may be deposited because of evaporation of water in saline inland basins or lagoons, when 37% of water has been evaporated.

According to the 'solar diagram' of Kurnakov, the order of crystallization is (i) Gypsum, (ii) Halite, (iii) Epsomite, (iv) Hexahydrite, (v) Carnalite and (vi) Bischofite.

It is suggested that anhydrite is deposited from water at a higher temperature than gypsum.

Mode of occurrence. The important commercial deposits of gypsum are those of rock-gypsum which occurs as beds with sedimentary rocks and are deposited from the solution by the evaporation of sea-water. Gypsite and albaster occur as beds or lenses and are formed by the evaporation of sea-water. 'Salenite' and 'satinspar' occurs as beds and lenses and are due to crystallization from solution.

Distribution in India. The most important sources of gypsum are in the state of Rajasthan. They are confined to the Tertiary rock-formations of Jodhpur region.

(*a*) **Rajasthan.** Beds of gypsum, half to two metres thick occur at several places around the Great Indian Desert of Rajasthan, particularly in the districts of Bikaner, (Jamser deposit), Jodhpur (Nagaur deposit), Barmar and Jaisalmer districts.

(*b*) **Tamilnadu.** In Tiruchirapolly district, where gypsum occurs as thin irregular veins in the clays and limestones of the Uttatur and Trichinopoly stages of the cretaceous system.

(*c*) **Jammu-Kashmir.** In the district of Barmula and Doda, rich deposits occur as lenticular bands in the pre-cambrian Salakhal schist or with nummulitic-limestones of Eocene age.

(*d*) **Himachal Pradesh.** Deposits, associated with Krol-limestone and dolomite and also with the Subathu series ; reported from Chamba, Mahasu and Sirmur district.

Besides the above, gypsum deposits also found to occur in the states of Gujarat, Uttar Pradesh, West Bengal, Madhya Pradesh, Andhra Pradesh etc.

Uses. It is used as a building material for the manufacture of cement ; for the production of various types of plasters, in pottery and statutary industries.

It is used as a fertilizer and also used in the paint, rubber and paper-industries, as well as in the manufacture of 'Plaster of Paris'.

CHAPTER 75

MAGNESITE

Magnesium does not occur free in nature. It is the lightest metal known and is found in a large number of minerals. The magnesium minerals of economic importance are

Magnesite	$MgCO_3$, MgO 47·8%
Dolomite	$MgCO_5$, $CaCO_3$
Brucite	$Mg(OH)_2$
Periclase	MgO
Serpentine	$H_4Mg_3Si_2O_9$
Olivine	Mg_2SiO_4
Carnalite	KCl, $MgCl_2(6H_2O)$
Spinel	$MgAl_2O_4$

Among the above minerals, magnesite is the only important ore-mineral of magnesium. There are two main varieties of natural magnesite (*i*) crystalline or spathic and (*ii*) amorphous (cryptocrystalline) or massive. In general the amorphous variety, though less common in occurrence is purer than the crystalline variety.

Origin :

1. By residual concentration.

2. It is generally believed as Sir Thomas Holland puts it that magnesite was formed by the action of superheated CO_2 and H_2O which were derived from magma. The superheated CO_2 and H_2O acted on the peridotite, dunite and serpentine converting the magnesium silicate to magnesite. It is also believed that SiO_2 which gets liberated occurs as chalcedony or quartz within magnesite.

$$H_4Mg_3SiO_9 + H_2O + CO_2 \rightarrow MgCO_3 + SiO_2 + H_2O$$

Serpenite   Magnesite

Mode of occurrence. Magnesite occurs as irregular veins and fracture zones in serpentine masses.

Distribution in India. In India, the principal magnesite deposit are found in Tamil Nadu (chalk hills and adjacent areas in Salem district and in Tiruchirapalli district). The other occurrences are

(*a*) **Karnataka.** Dodkanya and Dodkatur areas in Mysore and Hassan district.

(*b*) **Uttar Pradesh.** Someshwar and Bageshwar areas in Almora district.

Economic uses :

(*i*) Powder for flash lights, photography and fire works.

(*ii*) Alloys used in air planes.

(*iii*) Metal as deoxidiser and desulphurizer of Ni.

(*iv*) As refractory material, like furnace lining etc.

CHAPTER 76

KYANITE

It is a member of the aluminium-silicate group of minerals, where three minerals like andalusite, sillimanite and Kyanite have been included. All these are characteristic metamorphic minerals, having the chemical composition—$Al_2O_3SiO_2$ where Al_2O_3 constitutes about 62 93% of the mineral. But they differ only in the their crystallographic characters and other physical properties.

Origin. Kyanite is believed to have been derived from the argillaceous rocks metamorphosed under moderate temperature and high stress. It is a characteristic mineral of regional metamorphic rocks of medium grade and is absent from normal contact aureoles.

Mode of occurrence. It occurs as disseminated crystals in schists, quartzites and gneisses produced by regional metamorphism and also as surface deposits containing pebbles and boulders.

Distribution. Important deposits of Kyanite occur in the Singhbhum district of Bihar, along a belt 80 miles in length, stretching east along the western part of Seraikela, through part of northern Singhbhum and Kharsawan into Dhalbhum. The Lapsa-Buru Kyanite deposit, which is the largest Kyanite deposit in the world, is situated in this belt. Along this belt, kyanite occurs in the form of boulders and pebbles. The country rock is mica-schist. According to Dunn, kyanite rocks represent a smaller area in the entire belt which were even more basic than a normal clay and which was most probably a bauxite clay and no free quartz had been formed, when the process of metamorphism converted the clay into kyanite.

Kyanite deposits have also been found in Purulia district of West Bengal and in Ranchi district of Bihar.

Kyanite of a semi-precious variety is known to occur in association with calcite in the hills of Narnaul, Patiala and is used in the local market as a gem-stone, known as 'brug'.

Uses. Kyanite after being calcined and grained to powder is used as a refractory material, as well as in ceramic industries. Transparent kyanite is used as 'gem-stone'.

CHAPTER 77

DIAMOND

Diamond is a crystalline modification of pure 'carbon'. Depending on their crystallisation, the transparency of the crystals and the presence of inclusions, there is a number of varieties of diamond, which are as follows :

(*i*) **Diamond (proper).** Perfect crystals, transparent and are used as gem-stones.

(*ii*) **Bort.** Imperfectly crystallised diamonds with inclusions.

(*iii*) **Ballas.** Spheroid aggregates with a radiated structure.

(*iv*) **Carbonado.** Also called as black-diamond, which is fine-grained, compact and opaque.

Origin. From the study of primary deposits of diamonds, it has been established that they are formed in gas-saturated ultra-basic magma. This magma is capable of dissolving great quantities of carbon, which is the source from which diamonds crystallise. While some scientists believe that diamond crystallises in the magma chamber before the eruption of magma, others believe that diamonds are formed when the magma rises to the surface and rapidly cools in the channels and pipes, which serve as conduits.

Mode of occurrence :

(*a*) Diamond primarily occurs in ultrabasic intrusive igneous rocks like peridotite and kimberlite, in a disseminated manner. The mother rock usually occurs as pipes, necks or dykes. Altered peridotites, in which diamond occurs as crystals very sparsely disseminated is called '*Blue-ground*'.

(*b*) Diamond placers occur more frequently as eluvial placer and alluvial placers. A classic example of this variety occurs in Panna.

(*c*) It also occurs in stratified conglomerates. In the Vindhyan formations as well as in the Kurnool formations, diamond occurs as pebbles in the sandstones and conglomerate. Example : Panna (Madhya Pradesh).

Distribution in India :

1.　In Andhra Pradesh, diamonds are found to be associated with a bed of conglomerate at the base of Kurnool series (Vindhyan system). The best known locality is Wazrakarur in Anantapur distinct, where good diamonds are found to occur.

2.　In Madhya Pradesh, especially in Panna, diamond occurs in four district geological formations, *viz.*,

(*a*) agglomeratic tuff of Majhgawan Pipe (Post-Kaimur in age),

(*b*) conglomerate deposits,

(*c*) deep-alluvial deposits, and

(*d*) shallow-alluvial deposits.

3.　In Orissa, diamonds have been found in the Mahanadi alluvium at Hirakud in Sambalpur district.

Uses. The refractive-index of diamond is high (2·41 to 2·43), this alongwith its hardness make it economically most important. Excepting its use as gem-stone, it is also used as an abrasive. The principal industrial uses are in diamond-drill, in diamond tipped tools for truing and dressing abrasive wheels, boring and cutting metals, hard alloys, bakelite, hard rubber etc.

CHAPTER 78

COAL

Coal is one of the principal mineral fuels. As defined by Stutzer and Noe 'coal is a combustible rock which had its origin in the accumulation and partial decomposition of vegetation. Palaeo-botanists have shown conclusively that coal had been formed usually from land plants.

Composition. Chemically coals are composed of organic and mineral matter. Their organic mass consists of carbon (60 to 90%), hydrogen (1 to 12%), oxygen (2 to 20%), nitrogen (1 to 3%) and slight amounts of sulphur and phosphorus. The proportion of these elements progressively varies with the advance of coalification process starting from plant material (with the carbon content steadily increasing and hydrogen, oxygen contents decreasing). The chemical composition of coal is determined either by the 'Proximate Analysis' or the 'Ultimate Analysis'. A general evaluation of the quality of coal can be made on the basis of the data furnished by the proximate analysis regarding moisture content, volatile matter content, ash percentage, fixed carbon, heating value of coal etc.

Fixed carbon is the value obtained by subtracting the sum of the percentages of moisture, volatile matter and ash from 100, *i.e.*..

$$\text{Fixed carbon percentage} = \left[\begin{array}{l} 100 - (\text{Moisture percentage} \\ + \text{Percentage of volatile} \\ + \text{Ash percentage}) \end{array} \right]$$

$$\text{Fuel ratio} = \frac{\text{Fixed carbon}}{\text{Volatile content}}.$$

The heating value or calorific value of coal represents the amount of heat liberated by the complete combustion of a unit weight of coal. The value is represented in two standards of unit as (*i*) British, Thermal Unit (B.Th.U.) and (*ii*) Calorie.

Origin. It has been established that 'coal had its origin in the accumulation of vegetal matter, which has been subjected to a variety of geological processes bringing about marked changes in the physical and chemical composition. The changes are revealed by the gradual darkening of colour, increase in compactness, hardness and carbon content and decrease in moisture and volatiles.

Two views have been advanced to explain the origin of coal, viz., (*a*) Growth *in-situ* theory, (*b*) Drift theory.

Growth in-situ theory. This theory states that the coal-vegetation was fossilized practically on the site of growth, either due to tectonic movement or due to some other reasons. Evidences in support of this theory are as follows :

1. A vast amount of plant materials is accumulated '*in-situ*' in the present day swamps.
2. Many fossilized tree-trunks are found in erect positions with their roots firmly fixed in the underclays that lie beneath the coal seams.
3. A comparatively pure state of coal-seams indicates that the material accumulated without getting mixed with adventitious material and had not been transported along with sediments.

(*b*) **Drift-theory.** The drift theory is however strongly held by some geologists, which states that the coal-seams have been formed as a result of drifting and subsequent accumulation of the plant bodies away from their place of growth.

The evidences in support of this theory may be stated as follow :

1. No underclays, representing the soil at the root are associated with the coal.
2. Stems with roots in upright position are not found.
3. Beds of coals are observed to branch out, which is the characteristic only of drift matter.
4. Coal-seams occur in association with sedimentary rocks and itself behave like a stratified sedimentary rock.
5. Presence of 'channel sands' indicate criss-cross movement of water through swamps.

Formation of coal. The process of formation of coal is complex and involves both bacteriological and physical agencies. According to A.M. Bateman, the following things are essential for the formation of coal :

(*a*) **Source materials.** Plants and trees are the chief source material.

(*b*) **Places and conditions of accumulation.** The extensive distribution of individual coal seams implies swamp-accumulation, on broad delta and coastal plain areas, on broad interior basin low lands that have been base levelled etc. Thus the coal bearing horizon should be a basin like structure where the area should be swampy naturally.

(c) **Climatic condition.** The favourable climatic conditions are

(i) mild-temperate to sub-tropical climate,

(ii) with moderate to heavy rain-fall, well distributed throughout the year.

Stages of formation. There are two stages of formation of coal as

(i) Bio-chemical stage and the process is known as humification or Peat-forming stage.

(ii) Geo-chemical stage where the process of coalification takes place.

(i) **Humification process.** The changes brought about in the plant-debris during this process are due to the decay and decomposition of the substances like resins, lignins, proteins, cellulose etc. present in plants. These changes are brought about by the activity of bacteria and other micro-organisms which ideally thrive in swampy conditions. This process is also called fermentation and the result is the formation of a porus, fibrous and friable mass called 'Peat'.

A necessary step in the process of humification is that the decay be arrested before complete destruction ensues so that the residue can accumulate. This is accompolished by means of decay promoting bacteria that makes the stagnant water toxic, which prevent further decay of vegetable tissues and permits their preserval and accumulation.

(ii) **Coalification.** Peat once formed, under the prevailing conditions at depth, in the earth's crust and due to various geological factors, is transformed through various stages to coal. Thus peat is first converted to lignite, lignite to bituminous coal and bituminous coal into anthracite. This series of peat→lignite→bituminous→anthracite is called coalification. The rank of the coal increases at a place with depth.

Mode of occurrence. Coal occurs as a sedimentary rock in association with sandstone, carbonaceous shale and occasionally fireclay in a regular succession and with repetitions. Tertiary coal, in certain cases, found to occur as *in-situ* deposits. But Gondwana coal occurs as drifted deposits. Igneous intrusions in the form of dykes and sills are present in the coal seams. Generally the intrusives are of mica-peridotite, lamprophyre and basic dolerites.

Distribution of Indian coal. Coals of India belong to two principal geological periods

(*i*) The Lower-Gondwana Coals of Permian age, and

(*ii*) Tertiary coals of Eocene to Miocene age.

The greatest period of coal-formation, in India, is the Permian. The important coal-bearing formations are collectively known as Damudas and belong to the Lower-Gondwana system. The Lower-Gondwana coals account for more than 98% of the annual production of coal, which are generally of Bituminous-rank ; whereas in Tertiary coal-fields lignite predominates.

Gondwana coals. The Gondwana coals are largely confined to the river valleys like the Damodar, Mahanadi, Godavari etc. The workable coal seams are confined to the Damuda group of the Lower Gondwana, wherein they occur in two main horizons, (*a*) the Barakar measures of the lower-permain age and (*b*) the Raniganj measures of the Upper Permain age. The coal seams of the Barakar measures are more important because they are of better quality and occur in all the fields, whereas coal seams of Raniganj measures occur principally in the Raniganj coal-field only.

Barakar coals (of the Jharia coal-field) possess low moisture, low volatile, high fixed carbon, high ash, low sulphur and low phosphorous content. In comparison to this the Raniganj coals contain high moisture (3 to 10%), high volatiles, medium fixed carbon, medium ash, low sulphur and low phosphorus contents. While the Barakar coals are good coking and steam coals, the Raniganj coals are poorly coking but excellent steam coals.

Amongest the important lower-Gondwana coal-fields of India, mention may be made of

(1) Raniganj coal-fields of West-Bengal.

(2) The Jharia, Giridih and Bokaro coal-fields of Bihar.

(3) The Talchir coal-field of Orissa.

(4) The Umaria, Sohagpur, Mohapani, Korba and Pench-valley coal-fields of Madhya Pradesh.

(5) The Singreni coal fields of Hyderabad.

(*ii*) **Tertiary coals.** They principally occur in Assam, in the Himalayan foot-hills of Kashmir and in Rajsthan (Palna in Bikaner) in Eocene strata. Besides, lignite deposits are found to occur in South Arcot district of Tamil Nadu, in Kutch of Gujarat and also in the state of Kerala. The Neyveli lignite field of Tamil Nadu (which is of miocene age), is the largest lignite deposit of South India.

In India, coals of super-bituminous to anthracite variety occur in the Eonene formation of Kashmir along the Himalayan foot hills, as well as in the Lower Gondwana strata in the Eastern, Himalayan region.

Uses :

1. Coal is a primary source of heat and power (thermal power).
2. It is also used in the production of water gas.
3. In metallurgical operations, for the purpose of extraction of metals like iron, zinc etc.
4. Gasification of coal which leads to the production of coal gas, tar, coke etc.
5. Different types of varnish and germicides are also produced from coals.

Points to remember. Coals are composed of a number of bands. These bands are termed as vitrain, clarain durain, and fusain. These bands have different degree of lusture and friability.

CHAPTER 79

PETROLEUM

Petroleum is the term which means 'rock-oil'. This is one of the important 'mineral fuels', and is a complex mixture of hydrocarbon compounds with minor amounts of impurities, *viz.*, nitrogen, sulphur and oxygen. The liquid petroleum is called '*crude oil*', petroleum gas is called '*natural gas*' and the semi-solid to solid forms of petroleum are commonly known as asphalt, tar, pitch, bitumen etc.

Origin. A number of theories have been put forwarded for the origin of natural petroleum and depending upon the primary source materials the theories may be grouped as

(I) Inorganic theories.   (II) Organic theories.

(I) Inorganic theories :
(a) **Brethelot's alkaline-carbide theory.** According to him CO_2 might react with alkaline metals contained in the interior of the earth at high temperature with the formation of alkaline carbides. These on contact with water liberate acetylene which through subsequent processes of polymerisation and condensation forms petroleum.

(b) **Mendeleef's carbide theory.** It is believed that iron carbides within the earth on contact with percolating water form acetylene, which escapes through fissures to the overlying porous rocks and there condenses. This theory is based on laboratory experiment, but the presence of iron-carbide within the earth has not been established definitely.

(c) **Moissan's volcanic theory.** He suggests that volcanic explosions may be caused by the action of water on sub-terranean carbides, and may lead to the formation of petroleum.

(d) **Cosmic theory.** Taking into account the presence of small quantities of hydrocarbons occassionally in meteorites, Sokolov considers petroleum to be an original product resulting from the combination of carbon and hydrogen in the cosmic mass during the consolidation of the earth.

(II) Organic theory :

(*a*) This has been put forwarded by Engler. His theory is based on the fact that by destructive distillation of fish-blubber, a product similar to natural petroleum could be obtained.

According to him, petroleum is formed by a process of putrefaction of animal remains. Nitrogen thus eliminated and residual fats get converted into petroleum by earth's heat and pressure.

(*b*) **Vegetable origin theory.** On the basis of certain facts as deposits of petroleum found in close association with sedimentary deposits containing diatoms, seaweed, peat, lignite, coal, oil-shale of known vegetable origin, this theory has been propounded. Besides the above, other facts also support the theory :

(*i*) Oil and coal appear to have close relationship indicating a vegetable origin.

(*ii*) The large amount of methane in natural gas can be explained as produced by the decay of vegetable matter.

(*iii*) Oils closely resembling petroleum can be distilled from coal, lignite etc.

(*iv*) Microscopic vegetable remains have been noted in crude oil even though it is rare.

(*c*) **Animal-origin theory.** Since 95 percent of the oil-fields occur in marine sediments it is assumed that oil was formed from marine organisms buried in sediments. It has been suggested that bacterial action plays the most effective role in the conversion of organic material into oil. It is now commonly presumed that the primitive forms of life like diatoms, algae etc. which were mixed up and enclosed within the sediments in the sea-bed, were the primary source material for petroleum.

Mode of occurrence. Four pre-requisites are necessary for petroleum to accumulate in commercial quantities in an area :

1. The oil originates in a source bed, and a marine shale once a black-mud rich in organic compound is thought to be a common source rock.

2. The oil then migrates to permeable reservoir rocks and to do this, it may travel for long distances both vertically or horizontally. The source beds tend to lack the permeability necessary for profitable extraction of the oil.

3. A non-permeable layer must occur above the reservoir bed.

4. A favourable structure must exist.

Migration. Oil is found to occur in porous and permeable rocks like grits, sandstones, limestones and carbonate rocks which have served as reservoirs for oil and are called 'reservoir rocks'.

The migration of oil is thought to be caused due to compaction of sediments during diagnesis due to which currents of water are set up in the source rocks and the oil is squeezed alongwith water and enters into the porous and permeable rocks in the immediate vicinity.

Since oil is lighter than water, the oil tends to float on top of the water. If the sandstone unit was formed under marine water, it already contains salt water in its pore-spaces. The oil slowly moves up around the sand grains until it reaches the top of the sandstone unit. Gases that have been produced are lighter than the oil and tend to move to the top of the oil-accumulation. The greater the porosity, the greater the amount of oil that a reservoir rock can contain and the larger the pore space the greater the amount of oil.

Reservoir-trap. It holds the oil and gas in place so that they do not escape until released by drilling. It is also known as 'cap-rock' which may be an impervious shale, clay, dense limestone, well cemented-fine grained or shaly-sandstones that are effective cap rocks and they seal the reservoir trap.

The occurrences of petroleum deposits are classified into two broad-categories :

    (*a*) Surface occurrences   (*b*) Sub-surface occurrences.

(*a*) **Surface occurrence.** Petroleum occurs at the surface of the ground in the following ways :

    (*i*) seepages, spring and bitumen,
    (*ii*) mud flows and mud volcanoes,
    (*iii*) oil shales or kerogene shales.

(*b*) **Sub-surface occurrence.** Petroleum mostly occurs under the impermeable cap-rock of a reservoir. The barrier which helps in the accumulation of petroleum is called an oil-trap. Oil-traps are classified into the following three types :

    (*i*) Structural traps.
    (*ii*) Stratigraphic traps.
    (*iii*) Combination of stratigraphic and structural traps.

(*i*) **Structural traps.** These are caused by folding or rupture and displacement of the rock units. They include closed anticlines, domes, monoclines, terraces, synclines, faults, fissures, salt domes, igneous intrusion etc. The process of strata deformation may be compressional, gravitational, intrusional or rejuvenated uplifting.

(*ii*) **Stratigraphic traps.** These are formed by conditions of sedimentation in which lateral and vertical variations in thickness, texture and porosity of beds result. They include unconformities (angular unconformities are more effective), ancient shore-line sands, shoe-string sands, sandstone lenses and bars etc.

(*iii*) **Combination structural-stratigraphic traps.** Here are included these reservoirs where structural, stratigraphic and lithological features are significant in controlling the accumulation, migration and retention of oil and gas. They include both deformational as well as erosional features, for example bald headed structure, traps with burried hills etc.

Distribution in India. In India, deposits of petroleum and natural gas are associated with the belt of tertiary-rocks in Assam, Gujarat as well as in the offshore regions of Bombay High and in the Cauvery and Godavari deltaic areas.

Assam :

(*i*) *Digboi oil-field.* Where the Tipam sandstones of Miopliocene age is the oil-bearing formation.

(*ii*) *Nahorkatiya oil-field.* In the Brahmputra valley, where the oil-bearing formation is the Barail series of Oligocene age.

Besides the above, the other important oil-fields of Assam are 'Moran oil-field', 'Rudrasagar oil-field' and 'Lakwa-oil field'.

Gujarat :

(*i*) *Cambay-basin.* Where the main oil-bearing sand is of oligocene age. Here a majority of the wells are only gas producers.

The other oil-fields of importance are Kalol oil-field, Nawagam and Sanand oil-field.

(*ii*) *Ankleshwar oil-field.* This is the most important oil-field discovered so far in Gujarat. The producing sands are of 'Eocene-age'.

Bombay High. About 115 miles off Bombay, in the Arabian sea, a huge oil-deposit has been struck in limestone rocks of miocene age. This has proved to be the richest oil-deposit in the country. The deposit is estimated to be around 4 billion-tonnes.

Andhra Pradesh. In the off-shore regions of Tamil Nadu and Andhra Pradesh, a number of oil-deposits have been discovered by ONGC, recently. Among them the deposits of 'Nagapatinam' is the most important one, from the commercial view-point.

Besides the above, a number of discoveries have already been made in the states of Tripura, Gujarat, West Bengal etc.

Uses :

(*i*) The chief use of petroleum are as fuel particularly in transport operations.

(*ii*) The petro-chemical derivatives of petroleum have a wide range of uses in agricultural, industrial and medicinal industries.

(*iii*) It is also used for the purpose of generation of heat and power.

PART VIII

—STRATIGRAPHY
—PALAEONTOLOGY

CHAPTER 80

PRINCIPLES OF STRATIGRAPHY

Stratigraphy, also known as 'Historical Geology' is the study of the stratified rocks that aims at unravelling the geological history of the earth.

These studies are based on the following few principles :

1. **Law of superposition.** It states that "an overlying bed or lamina is younger than that underlying *it* ; under normal conditions".

2. **Doctrine of uniformitarianism.** "The study of the present is a key to the study of the past".

3. The geological events of the past are mostly indicated by the rock units from their lithological and palaeontological characteristics. Rock units are distinguished from each other by their colour, texture, composition, fossil contents etc. The smallest rock unit is a *bed* ; a number of individual beds together form a *formation* ; a number of formations together constitute what is known as a group ; finally an assemblage of associated groups constitute a *Super-group*.

4. **Facies.** When a particular system (with reference to the geological time-scale) in a country is represented by different kinds of rocks in its different localities, it is said to possess different facies.

For example, the cretaceous system of India is represented by geo-synclinal deposits in the type area of Spiti, marine transgressional deposits in the Coramondal coastal areas, fluviatile and estuarine deposit in Madhya Pradesh and igneous rocks of volcanic nature in the extrapeninsular region.

5. **Homotaxis.** The similarity of position of strata or system in a sequence not implying similarity of age is termed *homotaxis*.

When geological formations are equivalent in age but situated wide apart in different districts, countries or continents, they are said to be *homotaxial*.

6. Rock units formed at different places exactly at the same time are known as *synchronous-beds*. In these beds the same species of the fossils are found.

7. Contemporaneous. Those rock units which are formed near about the same time, marked by the presence of the same genera of fossils, they are said to be contemporaneous.

8. Geological time. *'Era'* is the largest grouping of periods of geologic time and each **era** covers many millions of years. The geologic *periods*, are formed of *epochs* and the epochs are in turn subdivided into *ages*.

The geologic time is well correlated with the stratigraphic units. For example, A geologic period represents a *system* ; rock units formed during an epoch is known as *series* ; and the rock units which are formed during an age are known as *stages*.

9. Stratigraphic correlation. It is of three types like-lithostratigraphic, bio-stratigraphic and chronostratigraphic correlation.

The lithostratigraphic correlations are made by the following methods :

(*i*) Continuity of contacts between units.

(*ii*) Lithologic similarity.

(*iii*) Stratigraphic position of a unit in a sequence of strata.

(*iv*) Well logs.

(*v*) Structural characteristics.

The Bio-stratigraphic correlation involves the following methods :

(*i*) Stage of evolution of fauna.

(*ii*) Guide fossils.

(*iii*) Faunal resemblance.

(*iv*) Position in a bio-stratigraphic sequence.

Chrono-stratigraphic correlations are made with the help of the following methods :

(*i*) Quantitative chronology—radioactive dating methods.

(*ii*) Eustatic changes in sea level.

(*iii*) Palaeontology.

CHAPTER 81

CHRONOLOGICAL SUBDIVISIONS

Geologists have attempted in various ways to estimate the age of the layers of rocks and the geological history of the earth. The entire life span of earth is called the geological time. Since the geological formations of Western Europe were studied first by the pioneers in the field of stratigraphy, the chronological sequence of geological formations, which they had established there is accepted universally as the standard. Thus the geological time scale has developed and this scale within its limits gives us a pretty idea of the sequence of events in the history of earth.

The geological time scale has been broadly divided into six major divisions known as 'Eras'. There are evidences that each era was ended by widespread geological disturbances called revolutions. Each revolution includes changes like upheaval of the earth surface in certain regions and subsidence and ultimate submergence of some regions beneath the seas. In each of the eras, there were lesser events which were less widespread and less far reaching in their effects than the revolutions making the separation of eras. These events are called disturbances in the earth-crust. They also mark definite breaks in the geological and fossil records and divide an era into periods, epochs etc.

The various eras are :

1. Archaean or Azoic.

2. Pre-cambrian or Proterozoic or Algonkian.

3. Palaeozoic or Primary.

4. Mesozoic or Secondary.

5. Cainozoic or Tertiary.

6. Quaternary.

The geological time-scale with the evolution of life may be represented in a tabular form as follows :

Era	Period	Epoch	Evolutionary Change
Quaternary	Recent	—	Age of man
	Pleistocene	—	Extinction of large mammals ; age of man.
Cainozoic	Tertiary	Pliocene	Man evolving, mammals abundant.
		Miocene	First man-like apes.
		Oligocene	Appearance of modern mammals.
		Eocene	Diversification of placental mammals.
		Palaeocene	Evolutionary explosion of mammals.
Mesozoic	Cretaceous	—	Dinosaurs reached peak and became extinct. Modern birds common. Rise of flowering plants.
	Jurassic	—	Appearance of 1st toothed bird, dominance of dinosaurs. 1st flowering plant appears.
	Triassic	—	Extinction of primitive amphibians, transition of reptiles to mammals ; gymnosperms dominant.
Palaeozoic	Permian	—	Exinction of ammonites and trilobites ; dwindling of ancient plants.
	Carboniferous	—	Amphibians dominant on land, rise of insects, seed ferns and gymnosperms dominant.
	Devonian	—	First forest, first gymnosperm, diversification in fishes.

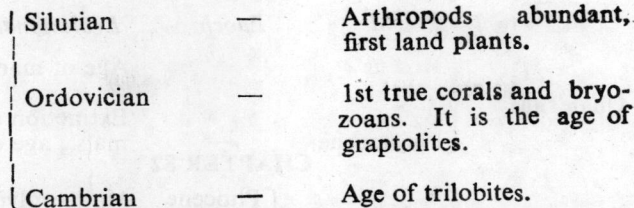

Silurian	—	Arthropods abundant, first land plants.
Ordovician	—	1st true corals and bryozoans. It is the age of graptolites.
Cambrian	—	Age of trilobites.

Pre-cambrian—Pre-cambrian—Life is still not prominant.

Archaean—Archean—Complete absence of living organisms.

Archaean rocks are mostly of igneous origin and are devoid of fossils. It has also been observed that fossils of simpler plants and animals represent older rocks and those having more complex forms are comparatively recently formed. The geological time scale represents the chronological subdivisions of the earth.

CHAPTER 82

INDIAN STRATIGRAPHY

For the study of the various stratigraphic units of India, references are usually made to the tripartite physiographic divisions of India. The *peninsula* is made up primarily of very ancient rocks of Archaean and Pre-cambrian age, as well as Deccan traps and the sedimentary formations of the Post-cambrian age. The *Indo-gangetic alluvial plains*, which is the second physiographic unit of India, belongs to the Quaternary era. The *extra-peninsula* made up primarily of sedimentary formations ranging in age from cambrian to Pleistocene.

ARCHAEAN GROUP

Recently, over 2500 million years age, has been more frequently used as the limit of the Archean era. The term 'Archaean' was introduced in the Indian Stratigraphy by J.D. Dana in 1872 to designate the geological formations older than Cambrian. According to Pitchamuthu (1965), the Archaean rocks of Peninsular India may be divided into two principal types, *i.e.*, the Schistose (Dharwars) and the Gneissose (Peninsular gneiss).

The stratigraphic relationship between the schistose and gneissos rocks of the Peninsular India has been the subject of much discussion ; while Bruce Foote, Oldham, Holland and Middlemiss believe that the gneissic rocks are forming the basement over which the sediments belonging to Dharwar Super-group are deposited ; Smeeth, Fermor believe that Dharwar supergroup constitutes the oldest geological formations in India. The basement problem is still controversial.

Age. From radiometric age dating the age of the Archaean rocks are estimated to be above 2500 million years.

Distribution. They cover 2/3rd of the peninsula. The areas occupied by the most ancient gneisses and schists are referred to as shields as they have remained virtually undisturbed and unaffected by changes. Peninsular India is a shield area.

1. **Peninsula.** South India, Madhya Pradesh, Rajasthan,

Singhbhum, Eastern Ghat ranges are some of the important type areas of Archaean rocks.

2. **Extra-peninsula.** The archaean gneisses and schists etc. are exposed along the entire length of the Himalayas from Kashmir to Burma, known as the 'Crystalline Axis'.

	Name
(i) In Kashmir-Hazara Area	Salkhala series.
(ii) In Kumaon-spiti area	Vaikrita system.
(iii) In Garhwal area	Jaunsar series.
(iv) In Simla area	Jutogh and Chail series.
(v) (a) In Nepal and Sikkim the Schistose rocks (Archaean)	Daling series.
(b) Gneissose rocks (Archaean)	Darjeeling series.
(vi) Bhutan	Buxa series.

Important features of Archaean rocks. They are all azoic, *i.e.*, there was no existence of life at the time of their formation. The close of the archaean era was marked by intense earth movements, as a consequence mountains and plateaus were formed and were acted upon by denuding agencies. Their subsequent subsidence and deposition of sediments from neighbouring land gave rise to *Eparchaean Unconformity.*

The important rock types are gneisses, schists, charnockites, khondalites, gondite, kodurite, pyroxenite, hyperite, quartzite, phyllite, marble, anorthosite etc.

In structure and composition, the archaean rocks are very complex and are therefore known as basement complex, gneissic complex etc.

1. **Archaeans of South India.** It was first studied by R.B. Foote. The archaean rocks of South India, are known as Dharwar system, and are best developed in Mysore and Southern Bombay. They are constituted by schists, gneisses and granites. The schistose rocks are isoclinally folded and the dip is towards east. The Regional Strike is N.N.W.—S.S E.

(i) **Classification.**

According to W.F. Smith :

(Younger) Chloritic Division→Chlorite and Mica schist, Quartzite, Marble, conglomerates etc.

(Older) Hornblendic Division→Hornblende schist, Calc-granulite, Hematite, Quartzite etc.

(*ii*) **Rama Rao.** Felsite and Porphyry dykes closepet granite (Bellary gneiss).

> Charnockites
> Norite dykes
> Hornblendic dykes
> Peninsular gneiss
> Champion gneiss

————UNCONFORMITY————

Upper Dharwar. (Sedimentary) : Conglomerates, calcareous silts, cherty and ferrugious silts and clay.

Middle Dharwar. Metamorphosed igneous and sedimentary rocks. The upper part of the sub-division consists of metamorphosed granitic rocks with gneissose structure. The lower part consists of micaceous genisses, schistose conglomerates, banded iron stones etc.

Lower Dharwar. Upper part is made up of highly crushed micaceous quartz-schists and gneisses, formed due to metamorphism of pre-existing acid volcanic rocks. Lower part is made up of Hornblende schist, green stones etc. the products of metamorphism of ancient dykes and sills.

————Base is not known————

Rama Rao has classified the exposures of Dharwarian rocks. into five distinct zones as

(*a*) **Eastern most zone.** Known as Kolar Schist Belt.

(*b*) **East central zone.** Composed of granulites, gneisses, schists of varying composition.

(*c*) **Central zones.** Composed of limestones, ferruginous rocks and metamorphosed igneous rocks.

(*d*) **West-central zone.** Composed of banded manganiferous and ferruginous rocks.

(*e*) **Western most zone.** Composed of hornblende schist, within which bands of hematite, quartzite occurs.

The Dharwarian rocks of south Bombay exhibit the lowest grade of metamorphism while those of the Southern Mysore are of the highest grade.

2. Archaeans of Madhya Pradesh. The areas where archaean rocks are developed characteristically in Madhya Pradesh and their characteristic features are as follows :

(*i*) In Bastar and adjacent areas the rocks are made up of schists and gneisses of both igneous and sedimentary origin, which have been subsequently intruded by igneous masses.

(*ii*) In Raipur, the rocks are mica-schist, phyllite, quartzites and Banded-hematite-quartzite etc.

(*iii*) In Bilaspur-Balaghat area, the oldest rocks occur towards the northern part and are known as *Sonawani series*, which is made up manganese ores, calc-gneisses, crystalline limestone, quartzites and schists.

The Sonawani series is overlain by the chilpighat series which is made up of trap rocks, grits, conglomerates, green stones etc.

The stratigraphic succession is as follows :

Amla granite

Porphyritic and Augen gneiss

Schistose and biotite gneiss

Chilpighat series ⎫
                  ⎬ Relationship is uncertain.
Sonawani series ⎭

It may noted that both Sonawani and Chilpighat series contain a manganese ore horizon and in part represent the Sausar series of Nagpur-Chindwara.

(*iv*) **Nagpur-Chindwara.** The chilpi rocks continue westwards and bifurcate, the southern strip occupying parts of Nagpur and Bhandara districts are known ar *Sakoli series.* The northern strip goes into Nagpur-Chhindwara, known as *Sausar series.*

Rocks of this series are granulites, marbles, schists and manganiferous rocks.

The Sausar series is divided into nine stages of which five are well developed in Nagpur area and four in Chhindwara area The rocks of the Sausar series generally dip to the South and S.S E. The stratigraphic succession is as follows :

Sapghota stage

Sitapar stage

Bichua stage

Junawani stage

Chorbaoli stage

Mansar stage—Gondites, Manganese ore etc.

Lohangi stage—Marbles and Manganese ores.
Utekata stage

Kadbikhera stage.

(v) **Nagpur-Bhandara area.** This is the Sakoli series. The rocks belonging to this series are less metamorphosed than the Sausar series. The rocks generally dip to the NNW and are made up of quartzites, dolomites, amphibolites, schists, phyllites, conglomerates etc.

The stratigraphic succession is as follows :

Quartz-dolerite

Granites and pegmatites

Sakoli series
Crushed quartzite, phyllite and slates.
Hematite-sericite-quartzites.

Chlorite-muscovite schist with chloritoid, jaspilite, phyllite, chloritic hornblende schist.
Amphibolites, dolomites, schists and quartzites.

The lower part of the sequence of metamorphic rocks is to some extent similar to Chorbaoli, Bichua and Sitapar stages of the Sausar series.

3. **Archaeans of Rajasthan.** Archaean rocks of Rajasthan occur

(a) in the Aravalli range region in the form of a part of the very large synclinorium ;

(b) in the plains lying to the east of Aravalli ranges and separated from the Vindhyan country further east by a great fault, known as 'Great Boundary Fault', which strikes approximately N.E.-S.W. ;

(c) in the desert regions of Jodhpur and Marwar to the west.

Important features. The archaean and other pre-cambrian rocks have together formed the conspicuous Aravalli ranges, which traverses the country along a N.E -S.W. direction.

Special attention is attached to the Aravalli mountains where the Dharwars were laid down in an ancient basin upheaved by orgen forces towards the close of the Archaean era.

The archaean stratigraphy of Rajasthan has been studied and worked out by C.A. Hacket, C.S. Middlemiss, A.M. Heron and many others. The stratigraphic succession is as follows :

Pre-cambrian——Delhi system.

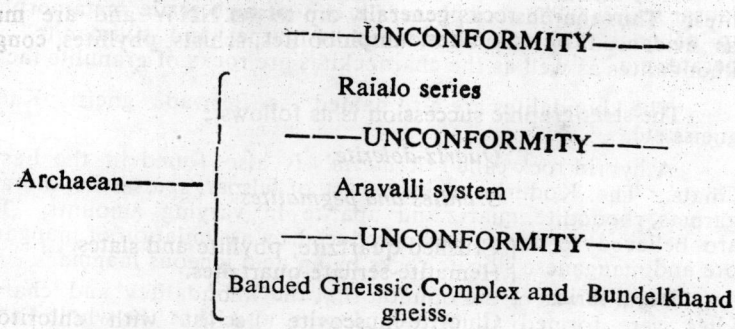

———UNCONFORMITY———

Raialo series

———UNCONFORMITY———

Archaean——— Aravalli system

———UNCONFORMITY———

Banded Gneissic Complex and Bundelkhand gneiss.

Bundelkhand gneiss. It is also called 'Berach granite' and may possibly represent the granitic portion of Banded Gneissic Complex. The exact relation between these two could not be established directly since their junction is concealed by the overlying Aravalli system.

Aravalli system. They are made up of mostly argillaceous materials. They show increasing metamorphism from east to west. The basal beds are arenaceous which are overlain by shales, phyllites and the associated volcanic rocks of basic composition.

(*i*) Reddish quartzites forming uppermost part of the Aravalli system are known as *Ranthambhor quartzites*.

(*ii*) **Binota shales** On the eastern side of the Great Boundary Fault, the unmetamorphosed Aravalli rocks are known as Binota shales which are succeeded by '*Jiran sandstone*'.

(*iii*) **Gwalior series.** Arenaceous rocks around Gwalior which are of the same age and are similar to the unmetamorphosed Aravalli rocks are called Gwalior series. Here the grits are called *Khardeola Grit.*

Raialo series. It overlies the Aravalli system and is made up of limestones. In Chitor area it is represented by unmetamorphosed dolomite which are known as *Bhagwanpura limestone.* Stromatolites are found in it. It is believed that Raialo series belong to the period represented by Eparchaean Unconformity.

An outcrop of archaean rocks near Baroda is called *Champaner series*, which contains quartzites, conglomerates, phyllites, marbles.

4. Archaeans of Eastern ghats. The Eastern-ghats region between Bezwada and Cuttack, which attains the greatest width in the Ganjam, Cuttack tract, is composed of ridges trending in a N.E.-S.W. direction. This is also the regional strike of the rocks.

The hills are made up of gneisses, charnockites and khondalites. Thus the Eastern ghats is a belt of high grade metamorphism as evidenced by the abundanee of garnet and sillimanite. The khondalites as well as the charnockites are rocks of granulite facies.

The khondalites are also named as Bezwada gneiss, Kailash gneiss etc.

A hybrid rock called 'Kodurite' is also found in the Eastern Ghats. The Kodurite rocks consist of felspar, spessartite-andradite garnets, rhodonite, quartz and apatite in varying amounts. They are believed to have been produced by assimilation of manganese ore and manganese silicate rocks by an acid igneous magma.

Fermor was of the opinion that the khondalites and charnockites were formed at greater depths and that the whole Eastern Ghats belt was uplifted at a later period bringing those high grade metamorphic rocks to the surface.

Near Koraput, an igneous complex of alkali-gabbro, calc-alkali syenite and nepheline-syenites occurs concordant to the foliation of the gneissic rocks of the region.

5. ARCHAEANS OF SINGHBHUM

The archaean rocks of Singhbhum have been studied by V. Ball, H.C. Jones, J.A. Dunn and A.K. Dey. The archaeans of Southern Singhbhum are separated from those of the Northern Slnghbhum by a distinct thrust zone, running approximately East-West for about 100 miles.

The stratigraphic succession in Singhbhum as a whole is as follows :

Pre-cambrian	Kolhan series
	———UNCONFORMITY———
	Newer Dolerite
	Arkasani granophyre or Soda-granite
	Singhbhum granite
	Chota Nagpur gneiss and granite
Archaean	Ultrabasic rocks
	Lavas (Dalma traps and Dhanjori lavas)
	Dhanjori Stage
	———UNCONFORMITY———
	Iron ore series.

The rocks within the thrust zone are metamorphosed shales and sandy shales in the main with subordinate metavolcanics, *viz.*, chlorite schist amphibole-schist, mica-schist talc-schist etc. The pelitic materials have been extensively migmatised resulting in felspathic mica-schist, quartz-muscovite—biotite-schist and highly felspathic granophyric rocks known as Soda granite or Arkasani granophyre.

The iron-ore series is made up of conglomerates, sandstones, limestones, shales, banded-hematite-quartzites and basic lavas.

The thrust zone of Singhbhum has been mineralised to a great extent. Important deposits of apatite, magnetite and of copper ore are found to occur in this belt.

CHAPTER 83

PRE-CAMBRIAN GROUP

After the formation of the Archaean group of rocks, they were subsequently subjected to diastrophism, erosion and denudation, which led ultimately to the development of what is known as 'Eparchaean unconformity'. The pre-cambrian rocks lie above this unconformity. In India the pre-cambrian rocks are divided into two systems known as (1) Cuddapah system, (2) Vindhyan system. The rocks of the Cuddapah system are older and structurally more complicated than the younger Vindhyan rocks.

CUDDAPAH SYSTEM

Structure. The huge Cuddapah basin is more or less crescentic in shape with the concave side facing the east. Enormous thickness of the Cuddapah sediments (6000 metres) indicates that a slow and quiet submergence was in progress all through their deposition. The western side exhibits undisturbed sequence of rocks while the eastern side shows steep folding.

The name of the Cuddapah system has been derived from the Cuddapah basin of Andhra state, where it is best developed.

Lithology. The rocks of the system are devoid of fossils and include quartzites, indurated sandstones, slates, shales and limestones with some banded jaspers.

Contemporaneous volcanic activity prevailed on a large scale during the lower half of the system, the records of which are left in a series of bedded traps and tuff-beds.

The stratigraphic succession of the Cuddapah basin as given by W. King is as follows :

Kurnool system—Made up of sedimentary
    rocks of Vindhyan age.

————UNCONFORMITY————

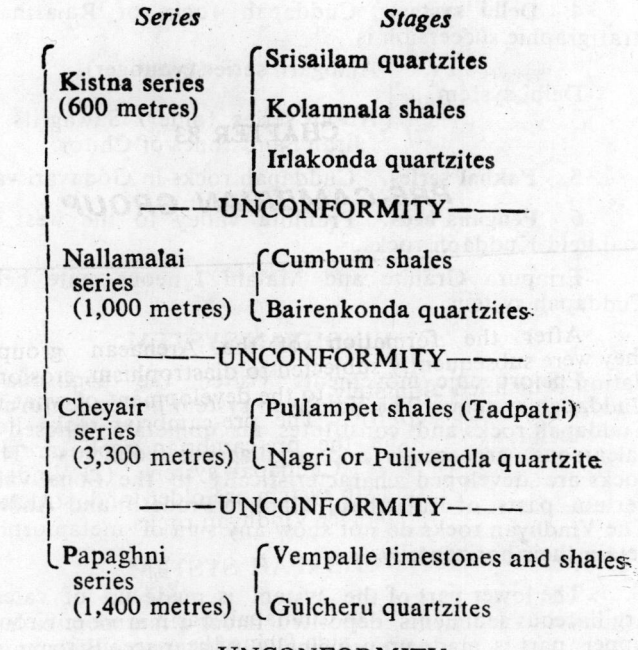

	Series	Stages
Cuddapah system	Kistna series (600 metres)	Srisailam quartzites Kolamnala shales Irlakonda quartzites
	———UNCONFORMITY———	
	Nallamalai series (1,000 metres)	Cumbum shales Bairenkonda quartzites
	———UNCONFORMITY———	
	Cheyair series (3,300 metres)	Pullampet shales (Tadpatri) Nagri or Pulivendla quartzite
	———UNCONFORMITY———	
	Papaghni series (1,400 metres)	Vempalle limestones and shales Gulcheru quartzites

———UNCONFORMITY———

Archaean→Schists and Gneisses.

The Cuddapah system is made up of alternate layers of quartzites and shales and there exists an unconformity between any two successive series of the system.

The Vempalle limestones have been intruded by dolerite and basaltic sills which range in thickness between wide limits, are possibly responsible for the development of deposits of barites, asbestos, etc. Limestone has been metamorphosed to marbles and on account of its being impure has developed minerals such as serpentine and talc. This kind of marble is known as *Ophicalcite'*.

Other Outcrops :

1. **Kaladgi series.** Cuddapah rocks occurring in South Maharashtra in the Bijapur district.

2. **Chandrapur series, Raipur series and Bijawar and Gwalior series** are rocks of Cuddapah age occurring in respective localities of Madhya Pradesh.

3. **Kolhan series.** Cuddapah rocks of Singhbhum.

4. Delhi system. Cuddapah rocks of Rajasthan and its tratigraphic succession is

Delhi system

⎧ Ajabgarh series (younger).

⎨

⎩ Alwar series (older) Sawagrits, shales and Jiran Sandstones of Chitor.

5. Pakhal series. Cuddapah rocks in Godavari valleys.

6. Pengana beds. Pranhita valley to the west of Wardha coal field Kuddaph rocks.

Erinpura Granite and Malani Igneous suite belong to the Cuddapah system.

VINDHYAN SYSTEM

Epeiorogenic movements visited the peninsula after the Cuddapah system. The Vindhyan system lies unconformably on the Cuddapah rocks and constitutes an unmetamorphosed column of calcareous, arenaceous and argillaceous sediments The Vindhyan rocks are developed characteristically in the Sone valley and in certain parts of Rajasthan, Madhya Pradesh and Andhra Pradesh. The Vindhyan rocks do not show any sign of metamorphism. They retain their horizontally.

The lower part of the system is made up of calcareous ard argillaceous sediments deposited under a marine environment. The upper part is made up principally of arenaceous rocks of Estuarine or Fluviatite origin. These rocks said to be devoid of fossils except a few doubttul organic remains found in Suket shales of Lower Vindhyan rocks in Rajasthan.

Two genera of primitive brachiopods 'Fermoria' and 'Krishnania' have so far been described in the Vindhyan rocks.

The stratigraphic succession of the Vindhyan system is as follows :

Series	Stages
Bhander series (450 metres)	Up. Bhander sandstones
	Sirbu shale;
	Lower Bhander Sandstones
	Bhander (or Nagode) limestones
	Ganurgarh shales

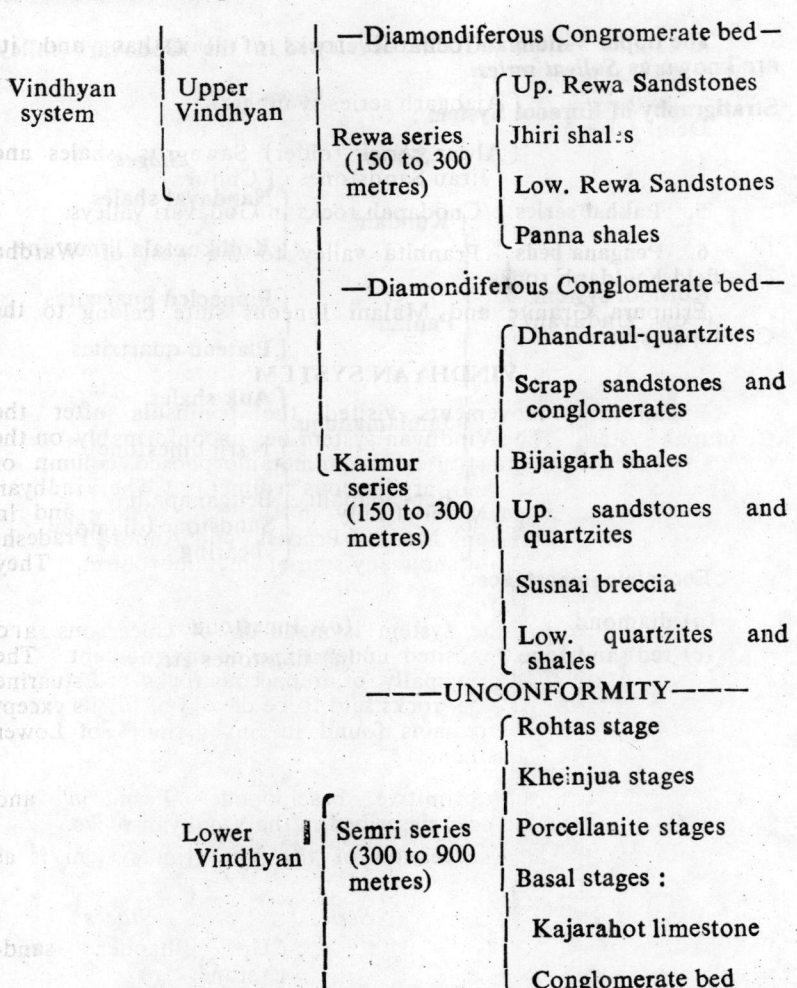

		—Diamondiferous Congromerate bed—	
Vindhyan system	Upper Vindhyan	Rewa series (150 to 300 metres)	Up. Rewa Sandstones
			Jhiri shales
			Low. Rewa Sandstones
			Panna shales
		—Diamondiferous Conglomerate bed—	
		Kaimur series (150 to 300 metres)	Dhandraul-quartzites
			Scrap sandstones and conglomerates
			Bijaigarh shales
			Up. sandstones and quartzites
			Susnai breccia
			Low. quartzites and shales
		———UNCONFORMITY———	
	Lower Vindhyan	Semri series (300 to 900 metres)	Rohtas stage
			Kheinjua stages
			Porcellanite stages
			Basal stages :
			Kajarahot limestone
			Conglomerate bed

The lower Vindhyan rocks have received various names according to the localities, where they are best developed. For example :

(*a*) **Bhima series.** Vindhyan rocks developed in the valley of Bhima, in Gulbarga and Bijapur district.

(*b*) **Kurnool system.** Vindhyan rocks developed in Kurnool and adjacent areas.

(*c*) In Rajasthan the lower Vindhyan rocks are represented by Jirohan limestone and breccia, the Suket shales, the Nimbhera shales and limestones. These together constitute the Malani series.

The upper Vindhyan rocks, developed in the Godavari valley are known as *Sullvai series*.

Stratigraphy of Kurnool System :

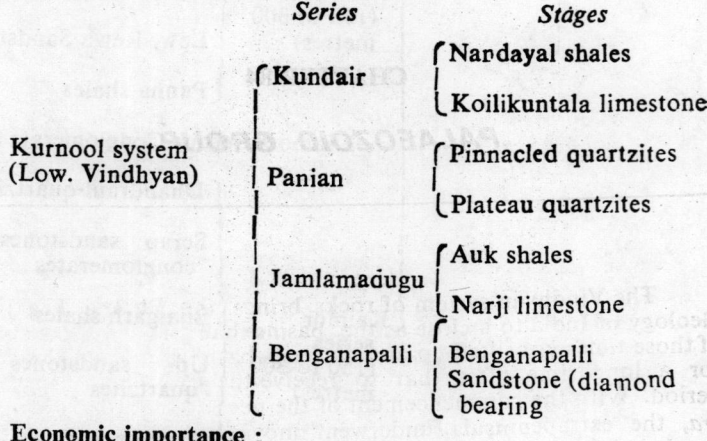

	Series	Stages
Kurnool system (Low. Vindhyan)	Kundair	Nandayal shales
		Koilikuntala limestone
	Panian	Pinnacled quartzites
		Plateau quartzites
	Jamlamadugu	Auk shales
		Narji limestone
	Benganapalli	Benganapalli Sandstone (diamond bearing

Economic importance :

(*a*) diamond   (*b*) limestone

(*c*) red sandstone   (*d*) flagstones etc.

CHAPTER 84

PALAEOZOIC GROUP

The Vindhyan system of rocks brings the Purana Era in the Geology of India to a close as the basins that received the sediments of those times got silted up and no fresh basin came into existence for a long time after that to receive the sediments of succeeding period. With the commencement of the new era called the *Dravidian era*, the extrapeninsula underwent movements of subsidence and developed marine conditions in certain restricted areas.

The geological formations developed during the palaeozoic Era are generally found to contain fossils. Therefore, in this respect, these rocks differ radically from the unfossiliferous formations of Pre-cambrian and Archaean ages.

Palaeozoic rocks formed under marine environment and containing distinct remains of organisms are developed characteristically in the extra-peninsula. In the peninsula, the palaeozoic group is represented by :

(*i*) Talchir and Damuda series of Permo-carboniferous age.

(*ii*) The marine lower permian rocks near Umaria in M.P.

(*iii*) A portion of the Vindhyan sequence which is possibly of cambrian age.

Salt Range. [West Pakistan].

In this area, the sequence commences with 'Cambrian-rocks' and so far no distinct pre-cambrian rock has been found to form the basement. Cambrian rocks overlain directly by the glacial boulder beds of Up. carboniferous age.

Fauna. Trilobites, Brachiopods, Foraminifera, Sponges, Echinoderms, Gastropods and Pelecypods.

Stratigraphic sequence :

Permian   Productus limestone series.

Low. Permian   ⎰ Olive series and Specked sandstones
        to   ⎱ series
Up. Carboniferous   ⎰ Glacial boulder bed.

———UNCONFORMITY———

⎰ Salt pseudomorphshale

⎪ Magnesian sandstone

⎪ Neobolous beds

Cambrian   ⎪ Purple sandstones

⎪   ———UNCONFORMITY———

⎩   Saline series.

The saline series is made up of salt marls. They are devoid
bedding planes. The top part contains, gypsum dolomite and oil
shale and also altered lava flows known as *Khewra trap*. Massive
gypsum are called 'Kalabagh Diamonds'. The age of the saline
series is still controversial.

The productus limestone series is subdivided into three parts as

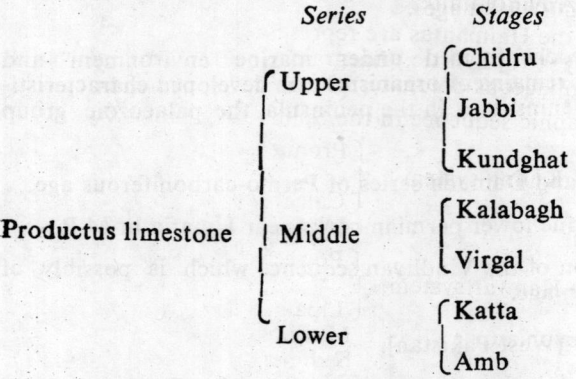

	Series	Stages
	⎰ Upper	⎰ Chidru
		⎪ Jabbi
		⎩ Kundghat
Productus limestone	⎰ Middle	⎰ Kalabagh
		⎩ Virgal
	⎩ Lower	⎰ Katta
		⎩ Amb

PALAEOZOIC OF KASHMIR

A complete succession of palaeozoic rocks exist in Kashmir and
adjacent areas and appear to have been folded conspicuously in the
form of synclines and anticlines. The cambrian fauna of Kashmir
differs from that of the Salt range and Spiti areas.

The stratigraphic succession is as follows :

Middle and Up. permian	Zewan beds
Up. Carboniferous to Low. Permian	{ Gangamopteris beds Panjal traps Agglomeratic slates
Carboniferous	{ Fenestella shales Syringothyris limestones
Up. Silurian and Devonian	Muth quartzites.
Silurian	Arenaceous shales and limestone.
Ordovician	Limestones and slates.
Cambrian	Clays, limestone, quartzites.

In Hazara area, unfossiliferous palaeozoic rocks made up of conglomerates, quartzites, phyllites etc. together form what is known as Tanwal series. It may range in age from cambrian to Up. carboniferous. It is overlain by the glacial Tanakki conglomerates of Permo-carboniferous age.

SPITI

The Haimanta system of Spiti overlies the Vaikrita system of Pre-cambrian age. The lower Haimantas are greenish phyllites, shales and thick bedded quartzites. The Mid. Haimantas are red and pink quartzites and shales. The Up. Haimantas, also known as *Parahio series*, are mainly grey and green quartzites, slates, shales with thin beds of grey dolomites.

In Kumaon the Haimantas are represented by Garbyang series consists of slaty to phyllitic, finegrained sandstone, calcareous sandstone and argillaceous dolomite.

The stratigraphic sequence in the Spiti valley is as follows :

Permian	Kuling system	{ Productus shale Calcareous sandstone
Up. Carboniferous		Conglomerate bed
Carboniferous	Kanwar system	{ Po series Lipak series
Up. silurian to Devonian		Muth quartzite
Ordovician		Sedimentary rocks
Cambrian	Haimanta system	{ Upper Sedimentary Middle Rocks Lower In all the parts.

The upper part of Po series, contains a stage known as Thabo stage which is overlain by another stage *'fenestella shales'*.

CHAPTER 85

MESOZOIC GROUP

With the close of the Mid-carboniferous, the Geology of India has come to the threshold of the last, the longest and the most eventful era called the 'Aryan Era', extending from the Up. carboniferous to the recent period. The important events occurred in chronological order are as follows :

1. The Himalayan Region was converted during the upper carboniferous time into a vast geosynclinal sea. This sea is called the 'Tethys'.

2. The site of Kashmir Himalayas from Pirpanjal to Hazara in the Northwest and Ladakh in the Northeast was the scene of the most violent type of volcanic activity. The peninsula developed a number of trough faults which served as receptacles for a great thickness of fluviatile and lacustrine deposits to form the rocks of the Gondwana system.

3. The Hercynian or Variscan revolution took place at this time.

4. The hypothetical continent, the Gondwanaland developed fissures and its different parts began to drift apart.

5. Stupendous masses of basaltic lava welled out from the earths interior in different parts of the globe. The Deccan-trap in the peninsula is an instance of this kind.

6. The Alpine mountain building movement set in and gave rise to the Alpine system, the Rockies, the Andes etc.

7. The sub-continent of India assumed its present set up.

8. Large parts of the globe fell into the grip of a frigid climate and immense masses of snow and ice covered them during the pleistocene times. This is called the Pleistocene Ice-Age.

In India the mesozoic rocks generally lies conformably above the rocks of the palaeozoic age. Mesozoic rocks have been divided into three system as Triassic, Jurassic and Cretaceous.

CHAPTER 86

GONDWANA GROUP

Subsequent to the deposition and uplift of the Vindhyan rocks during the Pre-cambrian era, the peninsula witnessed no further deposition of sediments for a pretty long time. During the Up. carboniferous period, however there commenced a new cycle of sedimentation in interconnected inland basins of fluviatile and lacustrine origin. This new phase of deposition of sediments continued up to the end of the Jurassic period. These inland sediments of Up carboniferous to Jurassic age occupying a vast tract in the peninsula and together constitute the Gondwana Group or system (named after the Gond Kingdom of M.P., where they were first studied by H.B. Medlicott in 1872). The Southern continents of the present day namely Australia, Antarctica, South America and India were during the Gondwana period united together to form one continuous stretch of land, known as Gondwana land.

Important features :

1. The deposition of Gondwana sediments commenced under glacial climatic conditions.

2. Then there prevailed a warm and humid climate during the rest of the Up. carboniferous and the whole of the Permian.

3. Throughout 'triassic', there prevailed a dry and arid climatic conditions.

4. During Jurassic again the country appears to have witnessed a more or less warm and humid climate.

5. Each individual cycle commenced with the deposition of coarse sand. The Gondwana rocks were subjected to tectonic disturbances during the Mid-triassic, Jurassic and Post-Eocene periods leading to the development of a number of faults in them.

They were also traversed by doleritic, lamprophyre dykes and sills which in the Damodar valley area are said to be genetically related to the *Rajmahal-traps*.

Classification. On the basis of the palaeontological, stratigraphical and lithological criteria, two types of classifications have been proposed for the rocks of this sequence :

(*a*)   The two-fold classification have been proposed by C.S. Fox, M S. Krishnan etc., in which they have divided the Gondwana rocks on the basis of floral characteristics into Lower Gondwana rocks characterised by Glossopteris flora and the Upper-Gondwana sediments marked by the advent of the Ptilophyllum flora.

These two divisions are separated from each other by slight unconformity.   This classification is as follows :

Subdivisions	Series	Stages	Ages
Upper Gondwana	Jabalpur	Umia	Low. Cretaceous
		Jabalpur	Up. Jurassic
	Rajmahal	Kota	Mid. Jurassic
		Rajmahal	Low. Jurassic
	Mahadeva	Maleri	Up. Triassic
		Pachmari	Up. Triassic
———UNCONFORMITY———			
	Panchet	Panchet	Low Triassic
———UNCONFORMITY———			
Lower Gondwana	Damuda	Raniganj	Up. permian
		Barren Measures	Mid. permian
		Barakar	Low permian
		Karharbari	Low. permian
	Talchir	Rikba	Up. Carboniferous
		Talchir	Up. Carboniferous
		Boulder bed	Up Carboniferous

(*b*)   The three-fold classification has been proposed by Feistmantel, Vredenburg, D.N. Wadia etc. on the basis of prevailing climatic conditions and faunal characteristics of the Gondwana era. The three subdivisions are as follows :

	Umia	Low. Cretaceous
	Jabalpur	Up. Jurassic
Upper Gondwana	Rajmahal	Up. Jurassic
	Kota	Lias
	Maleri	Keuper and Rhaetic
Mid. Gondwana	Mahadeva	Muschelkalk
	Panchet	Bunter
Low. Gondwana	Damuda	Up. permian
	Talchir	Up. Carboniferous

Inspite of several views for and against either of the classification, the modern trend is more in favour of the two-fold classification.

Important facts :

(*i*) Talchir stage is made up of greenish shales known as 'Needle shales', due to the characteristic weathering which they exhibit.

(*ii*) Damuda series contains coal seams of Permian age.

(*iii*) The Barakar stage is well developed in the Jharia coal field.

(*iv*) The Barren measures is devoid of fossils and contains nodules of clay-ironstone.

(*v*) Raniganj stage contains thick horizons of coal seams.

(*vi*) Panchet series was deposited in an arid climatic condition and is devoid of coal seams.

(*vii*) Himgiri beds in Mahanadi valley ; Pali beds of South-Rewa, Kamthi beds of Satpura in Madhya Pradesh belong to Damuda series.

(*viii*) Overlying the Damuda series of South Rewa is the Parsora stage.

(*ix*) Bap beds and Pokran beds in Rajasthan contain boulders of Vindhyan limestone and are assigned to the age of Talchir tillite.

(*x*) The Tiki beds of Rewa and Denwa and Bogra stages of Satpura are equivalent to the Maleri stage.

(*xi*) Rajmahal stage is principally made up of lava flows.

(*xii*) Dubrajpur Sandstone belongs to the Mahadeva series in the Raniganj hills.

(*xiii*) The upper part of the Kota stage is known as Chikiala.

(*xiv*) The Kota stage is well developed in Godavari valley.

(*xv*) Chaugan and Jabalpur stage belongs to Jabalpur series.

(*xvi*) The Barakar and Raniganj stages of the Damuda series constitute the most important coal-bearing horizons of the Gondwana succession.

(*xvii*) The rocks of the Gondwana system constitute a total thickness of 6,000 to 7,000 metres.

(*xviii*) The glaciated boulder bed called the Blaini Conglomerates of Simla Hills lying at the base is equivalent to the Talchirs.

(*xix*) The Tanaki Boulder Bed (Glacial) of Hazara is believed to be contemporaneous with that of the Talchir.

CHAPTER 87

TRIASSIC SYSTEM

The most complete section of the Trias is exposed in the Spiti-Kumaon belt of the Himalayas north to its main axis. The best known succession is exposed near Lilang, where the rocks are classified as part of the Lilang system.

Important features :

(*i*) The permian system of rocks like '*Productus Shales*' in Spiti is conformably overlain by the rocks of the Triassic system.

(*ii*) It is entirely marine in character.

(*iii*) The rocks of the system extend over enormous distances without any variation in their composition.

(*iv*) They consist of limestones and shales.

(*v*) Arid climatic condition and universal marine regression is characteristic.

(*vi*) In the type area of Spiti, the deposits are marine geosynclinal ; in the peninsula it is fresh water, fluviatile deposits.

(*vii*) On the basis of the fossil contents the system is divided into three series, *viz*., the lowest called the *Bunter*, the middle the *Muschelkalk* and the upper the *Keuper*.

(*viii*) The 'Lower trias' is 13 metre thick, the middle one is 120 metre thick and the upper trias is 1600 metre thick.

GEOLOGICAL SUCCESSION

Age	Series	Stages
Lower Jurassic and Upper Triassic	Kito (or Megalodon) limestones	Tagling stage Para stage

Upper Triassic (Keuper)	Noric	Quartzite series
		Monotis shale
		Coral limestone
		Juvavite beds
	Carnic	Tropites beds
		Grey beds
		Halobia beds
Mid. Triassic (Muschelkalk)	Ladinic	Daonella limestone
		Daonella shales
	Muschel-kalk	Up. Muschelkalk
		Low. Muschelkalk
		Nodular limestone
		Basal Muschelkalk
Low. Triassic (Bunter)		Hedenstroemia beds
		Meekoceras bed
		Ophiceras bed
		Otoceras bed

| Permian | Productus shale |

The Lower-triassic rocks are principally constituted by limestones. The otoceras beds which are the basal parts of the lower triassic sequence is named after the characteristic ammonite genus 'Otoceras'. The fauna of this bed is as follows :

Cephalopods. Otoceras woodwardi, otoceras draupadi, otoceras parbati.

Brachyopods. Rhynchonella.

Lamellibranchs. Pseudomonotis griesbachia.

The ophiceras bed contains the typical zone fossil 'Ophiceras Shakuntala'. The Meekoceras beds lies over the ophiceras beds and contain the remains of the characteristic cephaloped 'Meeko-ceras Varaha'. The Hedenstroemia beds which constitute the upper part of the lower triassic rocks are separated from the underlying rocks by an unfossiliferous zone of 1·5 metre thick. The characte-ristic fossils of these beds are Hadenstroemia mojsisovicsi, Xeno-discus, Flemingites etc.

The Middle-Triassic rocks are made up mainly of limestones, shales and shaly limestones and they lie conformably over lower triassic rocks. This series is highly fossiliferous. The Basal Muschel-kalk bed contains the characteristic fossil—Rhynchonella griesbach (Brachiopoda). The overlying Nodular limestones contain no¹ important fossils. The Lower Muschelkalk beds contain remains of ammonitic cephalopods like 'Sibirites Prahlada', Keyserlingites, Ceratite, Spiriferina, Rhynchonella etc. The Upper Muschelkalk limestones are highly fossiliferous containing remains of cephalopods like Ceratites, Ptychites rugifer and a few remains of Brachiopods.

Above the Muschelkalk stage lies the Ladinic stage. There is no noticeable change in the stratigraphical unit between Ladinic and Muschelkalk stage. The Daonella shales which constitute the lower part of the Ladinic stage contain important fossils like-Daonella indica, Ptychites geradi etc. The Daonella limestones which overlies the Daonella shales are consisting of fossils like Daonella indica, Joannites, Rhynchonella etc.

The Upper-Triassic system of beds are made up principally of limestones, shaly limestones, and quartzites that lies conformably above the older beds. The lower part is known as Carnic stage which is made up of three characteristic beds. The Halobia beds contain fossils like-Halobia Comata, Joannites etc. The Grey beds are characterised by the fossils like Joannites, Spiriferina, Rhyn-chonella etc. The overlying Tropites beds contain remains of 'Tropites subbullatus', Dielasma Julicum, Jovite spectabilis etc.

Above the carnic stage, there lies the Noric stage, having four important beds. The lower most beds of this stage are called Juvavite beds characterised by the ammonites-'Juvavite angulatus'. The overlying coral limestone beds contain remains of crinoids, corals and brachyopods. The characteristic fossils are Spiriferina griesbachi and Rhynchonella. The 'Monotis shales' lie above the coral limestones and are characterised by 'Monotis salinaria', Pecten, Lima, Rhynchonella etc. The quartzite beds are found above the Monotis shales and contain 'Spirigera-maniensis'.

The Kito (or Megalodon) limestones, lying above the quartzite beds, ranges in age from Rhaetic to Lias. It is unfossiferous in most of their parts and contain 'Megalodon Ladakhensis' the characteristic fossil. These limestones are generally divided into lower—'*Para stage*' and the upper-'*Tagling stage*'. The higher units within the limestone succession contains specimens of Spiriferina, Lima, Pecten etc. These fauna clearly suggests the upper triassic age for the 'Para-stage'. The Tagling stage' is unfossiliferous and is probably of Jurassic age.

JURASSIC SYSTEM

The Jurassic system was marked by extensive marine transgression and humid tropical climatic conditions. This system shows a more widely divergent facies of deposits in different parts of the country. In the Spiti area the deposits are marine geosynclinal ; in Kutch they are marine transgressional deposits, within the peninsula the deposits are freshwater, fluviatile deposits.

In the type area of Spiti in the Kumaon Himalayas the Triassic system is conformably overlain by the rocks of the Jurassic system. The junction between the two being so transitional that but for the change in their fossil contents, it would have been well nigh impossible to mark off one from the other.

The upper portion of this system is composed of shales indicating a shallowing of the sea floor towards the later part of the period, whereas the lower portion is dominantly composed of limestone.

In the peninsula, mesozoic rocks ranging in age from Mid. Jurassic to Low. Cretaceous are particularly well developed in Kutch, where they exhibit a complete succession and a total thickness of about 2,000 metres. But it is to be borne in mind that here as elsewhere such rocks are the result of what is known as marine-transgression or the invasion of sea on land possibly due to the subsidence of the peninsular block as a whole.

In Kutch, the Jurassic rocks constitute the oldest formation and at some places underlain by a few patches of rocks of pre-cambrian age. The Jurassic rocks in Kutch outcrop along three more or less parallel anticlinal ridges running approximately east to west. In Kutch, the Jurassic rocks are divisible into four series, *viz.*, Patcham series, Chari or Habo series, Katrol series and Umia series.

Umia series	1000 metres thick
Katrol series	400 metres thick
Chari or Habo series	450 metres thick
Patcham series	400 metres thick

In kutch, the underlying archaeans gneisses and schists are overlain by 2000 metres of sandstones, shales, and limestones. The geological succession of the Jurassic system in Kutch are as follows ;

Age	Series	Stages
		⎰ Marine sandstones
		⎪ Bhuj beds
		⎪ Umia plant beds
Up. Jurassic	⎱	⎪ Ukra beds
to	⎱ Umia	⎪ Barren sandstones and shales.
Low Cretaceous	⎰	⎪ Trigonia beds
		⎪ Ammonite beds
		⎱ Zamia shales
		⎰ Gajansar beds
		⎪ Up. Katrol
Up. Jurassic	Katrol	⎪ Mid. Katrol
		⎪ Low. Katrol
		⎱ Kantkote sandstone
		⎰ Dhosa oolite or Mehba oolite
		⎪ Atheleta beds
Middle to	⎱ Chari	⎪ Anceps beds
	⎱ or	⎪ Rehmani beds
Upper Jurassic	⎰ Habo	⎱ Macrocephalus beds
		⎰ Patcham coral beds
Middle Jurassic	Patcham	⎪ Patcham shell limestone
		⎱ Patcham basal limestone

The Jurassic succession in Kutch commences with the *Patcham series* which is made up principally of limestones, some sandstones and shales. The lowest beds, seen near Khera are known as '*Kuar Bet Beds*' constituted of clastic sandstones and yellow limestones.

The Patcham basal limestones contain the characteristic fossils 'Megateuthis' ; the overlying shell limestones contain pelecypods

(Trigonia and Corbula) and ammonites. The Patcham coral beds are characterised by the presence of remains of corals (stylina) and ammonites.

Above the Patcham series lies the *Chari series,* which takes its name from the Chari village. According to the fresh fossil collections made from the Habo dome, and since the rocks belonging to this formation are nowhere found in the vicinity of the village 'Chari' but are well exposed around the village 'Haboae' this series has been recently named as *'Habo series'.*

The oldest stage of Habo series is made up of shales and limestones and is known as *'Macrocephalus Beds'* characterised by the presence of 'Macrocephalites macrocephalus'. The upper part of the middle division of this stage contains a few layers of golden oolites constituting the Diadematus zone, which is a calcareous oolite and its grains are coated with thin films of Ferric oxide giving them a golden colour. This zone is named after the common ammonite 'Indocephalites diadematus'.

The overlying Rehmani beds contain remains of a number of ammonites-'Reineckeia rehmanni', Sivajiceras, Phylloceras and Lytoceras etc. The Anceps bed is characterised by the presence of the fossils 'Perisphinctes anceps'. The marls and gypseous shales overlying the anceps beds are called *'Atheleta Beds',* containing fossils like *'Peltoceras atheleta'.* The top most beds of the Chari series are the Dhosa oolites, composed of green and brown oolitic limestones. The important fossils of these beds are Dhosaites, Mayaites maya, Epimayaites polyphemus etc.

The *Katrol series* lies above the Chari series is composed of different types of sandstones and shales. The Kantkote sandstones is the oldest stage containing fossils like Epimayatis transience, Torquatisphinctes torquatus. The lower Katrols contain fossils Waagenia, Streblites etc. The red sandstones from the middle katrol horizon exhibit fossils like Virgato-sphinctes, Katrol.ecras, Waagenia etc. The upper Katrol rocks contain no fossils. The Gajansar beds are characterised by the presence of fossils like Belemnopsis geradi, phylloceras etc.

The Umia series lies above the Katrol series. The bottom most horizon 'Zamia shales' contain cycads and other plants. It is followed upwards by ammonite beds, which in turn are overlain by sandstones and conglomerates. It is subsequently overlain by Trigonia beds characterised by fossils like Trigonia ventricosa. Above it lies the unfossiliferous Ukra beds, but some plant remains are found occasionally. The Umia bed contains plant remains to a major extent like-Brachyphyllum, Williamsonia etc. The Bhuj beds are probably estuarine and contain ptilophyllum and other plants closely related to the Jabalpur series, of Gondwana system. The marine sandstones which occur above this stage is characterised by the presence of the fossils 'acanthoceras'.

Other Distribution :

1. In the Spiti valley, above the Kito Limestone, the upper Jurassic rocks are found to lie unconformably and are represented by the '*Sulcacutus beds*'. Above the Sulcacutus beds lies unconformably the 'Spiti-shales' which is made up of three stages, *viz.*, the lower is called '*Belemnite geradi beds*', the middle is the '*Chidamu beds* and the upper is the '*Lochambal beds*'.

The geological succession may be represented as follows :

Age	Series	Stages
Up. Jurassic	Spiti shales	⌈Lochambal beds │Chidamu beds ⌊Belemnites geradi beds
———UNCONFORMITY———		
Up. Jurassic	Sulcacutus beds	
———UNCONFORMITY———		
Low. Jurassic and Up. Cretaceous	Kito and Megalodon Limestone	⌈Tagling stage │ ⌊Para stage

CRETACEOUS SYSTEM

The cretaceous system as developed in India has a uniqueness of its own in that it has the most varied facies of deposits. No other geological system shows a more widely divergent facies of deposits, in the different parts of this country than the cretaceous and there are few which cover so extensive area of the country as the present system does in its varied form :

1. In the type area of Spiti, in the extra-peninsula region the deposits are marine geosynclinal.

2. The coromandal coastal strip received the marine transgressional deposits, while right in the heart of peninsula, there exists a chain of outcrops of marine cretaceous strata along the valley of Narmada which is also the result of the incursion of sea.

3. In Madhya Pradesh, the deposits are fluviatile and estuarine.

4. An igneous facies is represented in both its intrusive and extrusive phases by the records of a gigantic volcanic outburst in

the peninsula and by numerous intrusions of granites, gabbros and other plutonic rocks in many parts of the extra-peninsular region.

The heterogeneous constitution of the cretaceous is a part of the pre-valence of very diversified physical conditions in India at the time of its formation.

Distribution. (*a*) In the peninsula, (*b*) In the extra-peninsula.

(*a*) **In the peninsula** :

1. As we know, the peninsula as a whole continued to exist as a landmass after the Vindhyan-period and only a few patches of rocks of sedimentary origin of post-Vindhyan period were deposited along the coastal tracts during a few subsequent periods of marine transgression. The mesozoic succession in South India is represented by marine cretaceous rocks which are exposed in the Trichinopoly, Vridhachalam and Pondicherry districts.

Important features :

(*i*) The cretaceous rocks of South India rest upon a basement of Archaean gneisses and charnockites, and are sometimes fringed along their western margin, by thin strips of rocks of upper-Gondwana age.

(*ii*) Of the three patches, the one at Trichinopoly has the largest geographical extent, but the other two are much smaller.

(*iii*) These rocks appear to have been deposited as a result of the Universal Marine Transgression which occurred during the middle-cretaceous. This is known, as *Cenomanian Transgression.*

(*iv*) As it contains remains of a thousand extinct sea-animals, Sir T. Holland calls this series of beds '*A little museum of palaeozoology*'. The fossil fauna from these beds indicate that there are at least four phases of marine transgression during this period.

(*v*) The cretaceous rocks of South India, particularly in the type area of Trichinopoly exhibit a complete succession and a total thickness of more than 1000 metres, ranging in age from Aptian to Maestrichtian.

The geological succession is as follows :

Age	*Stage*
Danian	Niniyur formation

———UNCONFORMITY———

| Senonian to Maestrichtian | Ariyalur formation |

Turonian to Senonian	Trichinopoly formation

————UNCONFORMITY———

Up. Albian to Turonian	Uttatur formation

————UNCONFORMITY———

Aptian to Low. Albian	Dalmiapuram formation

————UNCONFORMITY———

Tripati and Pavulur formation.

The cretaceous succession in Trichinopoly commences with the 'Dalmiapuram stage', which was first recognised by Subbraman (1968) in the vicinity of Dalmiapuram, Bhatia and Jain (1969) after detailed study of Ostracods from this unit and stratigraphical observations made by them in the field suggested that these beds are Aptian to Albian in age. The rocks of this formation consists of grey shales with abundant pyrite and marcasite crystals and tiny flakes of muscovites. They contain large number of ammonites, smaller foraminifera and ostracods. The earliest marine transgression took place in Aptian times resulting in the deposition of Dalmiapuram formation.

The *Uttatur formations* are composed of limestones, clays and arenaceous rocks. This stage lies on charnockites for the greater part. The lower part of this stage is made up of limestone beds, coral reefs and clays containing characteristic fossils like Schloenbachia inflata, Turrilites, Acanthoceras, (all cephalopods) ; Rhynchonella (brachiopod), Arca, Lima, Pecten (pelecypods) etc. In addition, a few fragments of fishes, cycadeous woods, some ostracodes, also occur in this formation.

Uncoiled ammonites are characteristic of the Uttatur formation, which is formed during the second phase of marine transgression.

The *Trichinopoly stage* is a shallow marine formation consisting of sandstones, calcareous grits, occassional shales and limestones. The limestones of this formation is full of white shells of gastropods and pelecypods and is locally known as 'Trichinopoly Marble'. The Trichinopoly formation is distinguished from the underlying Uttatur formation by the presence of a large number of granite pebbles in the gravels and conglomerates. It has been formed during the third phase of marine transgression. The fossils are as follows :

Cephalopods. Placenticeras, desmoceras, holoceras etc.

Pelecypods. Trigonia, Spondylus, Corbula, Ostraea etc.

Gastropods. Cerethium, Turritella, Gosavia indica etc.

Corals. Trochosmilia, Astrocoenia etc.

The Ariyalur stage consists of mostly argillaceous sandstones and white sandstones. The fauna of this stage resembles that of the underlying Trichinopoly stage. The fossil assemblages of this formation is as follows :

Cephalopods. Schloenbachia, Nautilus, Baculites.

Gastropods. Cerithium, Nerita, Cypraea.

Pelecypods. Cardita, Nucula, Gryphaea.

Echinoids. Cidaris, Hemiasters.

Fishes. Ptychodus, Otodus.

The final phase of marine trangression took place in lower Maestrichtian times, during which beds of Ariyalur formations were deposited.

The *Niniyur formation* lies over the Ariyalur stage, and the ammonites have been found to disappear in this stage. The beds are made up of brown and grey sandstones, shales and arenaceous limestones. They contain the following fossils :

Cephalopods. Nautilus danicus.

Gastropods. Turritella.

Corals. Stylina.

Algae. Orioporella malaviae.

2. **Narmada valley.** Marine cretaceous rocks that developed in the Narmada valley covering an extensive area from Gwalior to Kathiawar are called 'Bagh beds'. They consist of sandstone, shale, marl, impure limestones and some quartzites. The lower part of the beds is arenaceous while the upper part is mainly calcareous. The stratigraphic position is as follows :

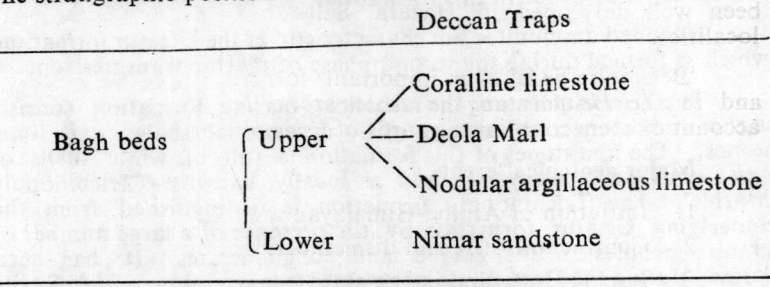

		Deccan Traps
Bagh beds	Upper	Coralline limestone
		Deola Marl
		Nodular argillaceous limestone
	Lower	Nimar sandstone
		Metamorphics

3. Madhya Pradesh :

Lameta beds. These rocks are of estuarine, lacustrine and partly fluviatile origin. They consist of silicified limestones, argilla-ceous sandstones etc. They lie beneath the Deccan traps and range in age from Up. cretaceous to Low. tertiary.

(B) Extrapeninsula :

1. Spiti. The Spiti shales of the Jurassic system are comformably overlain by the Giumal series of Low. cretaceous age. The succession is as follows :

Cretaceous	⎧ Flysh ⎪ Chikkim series ⎩ Giumal series
Up. Jurassic	Spiti shales

The Chikkim series overlies the Giumal series and in turn is overlain by Flysh deposits.

2. Kashmir. The cretaceous rocks are meagrely developed between Burzil and Deosahi plateau. Igneous activity during the cretaceous times in Kashmir shows itself as a great thickness of ash beds, tuff and agglomerate and bedded lavaflows of basic composition.

3. Sind and Beluchistan. Cretaceous rocks including limestone, sandstone and shales are found in the Sind and Beluchistan region. The sandstones are known as *Pab-sandstones* which are overlain by *Cardita-beaumonti beds*.

In Beluchistan, the limestones of this period are called 'Parh limestone'.

Besides the above, cretaceous rocks are also found to have been well developed in Hazara, Salt-range, Assam and adjoining localities.

'Belemnite beds' is an important formation in the Salt-range and in the Beluchistan the shales are called 'Belemnite shales' on account of their containing guards of *belemnites*.

Major geological events :

1. Initiation of Alpine-Himalayan mountain-building activity.
2. Continental drift and dismemberment of Gondwana land.
3. Igneous intrusion in the extra-peninsular India.
4. The volcanic activities in peninsular India.

CHAPTER 88

DECCAN TRAPS

Towards the end of the Mesozoic period, after the formation of the Bagh and lameta beds, the Indian peninsula was affected by intensive volcanic activity, due to which stupendous masses of lava and pyroclastic materials were ejected out, which covered a larger part of the peninsula in its southern, western and central parts and is of maximum extension next to the Archaeans. The lava flows occur in general in the form of beds, obliterating the previous topography and converting the countries into plateaus. Because of their tendency to form flat-topped plateau-like features and their basaltic composition, they are termed as plateau basalts. Their step-like or terraced appearance is suggestive of the name Deccan Traps to these volcanic formations.

General Characteristics :

(*i*) The eruption of volcanic materials was mostly through the fissures and the Deccan Traps are believed to be the result of fissure-type of eruption. But at a few places, like Girnar hills, Ranpur, Dhank, etc. The eruptions were of the central-type showing differentiated rocks of various characters.

(*ii*) **Thickness.** The lava-flows generally occur in the form of horizontal sheets ranging in thickness from seven metres to as much as thirty metres. It has a maximum thickness of about 3300 metres near Bombay.

(*iii*) **Areal extension.** The lava flows flooded several hundred thousand kilometres of Western, Southern and Central parts of the country covering an area of 320,000 square kilometers. They occupy a large part of Hyderabad, Bombay, M.P., Cutch, Kathiawar and Gujarat.

(*iv*) **Structure.** The traps show well-developed columnar jointing caused by tensile stresses, the result of contraction due to cooling. The columns are pretty long and polygonal in shape. The traps show a good deal of amygdaloidal structure.

The step-like appearances of the out-crops are because of differential composition and weathering.

(*v*) **Composition of texture.** The traps are essentially a basic rock of basaltic composition, and is a dark coloured or melanocratic rock. There are very little signs of differentiation but in some places like Pavagarh hills in Gujrat and the Girnar hills of Kathiawar differentiated rocks are also observed. According to Washington, the mineralogical composition of the trap-rocks are as follows :

Quartz	2 to 5%	Orthoclase	5 to 7%
Labradorite	40 to 50%	Pyroxene	30 to 40%
Iron-oxide	10 to 12%.		

(*vi*) **Inter-trappean beds.** As the volcanic eruption was not continuous, a number of gaps are found in the lava flows. The successive lava flows are commonly separated from one another by sedimentary beds, which were formed under water, containing valuable palaeontological and palaeobotanical data, throwing light on the history of periods of quiscenc which intervened between the volcanic outbursts.

According to Bose (1973), the Deccan volcanism was probably related temporarily to the Gondwana-rift and dispersal of the Indian block.

(*vii*) **Stratigraphic relations of the traps.** In other words, the stratigraphic classsfication of the deccan traps are as follows :

The Deccan Trap formation has been classified into lower, middle and upper divisions by different workers :

Upper Eocene	Nummulitic limestones of Surat and Broach.

———UNCONFORMITY———

Upper flow (450 m thick)	Flows with numerous inter-trappean sedimentary layers and volcanic ash beds. Exposed in parts of Bombay and Kathiawar. They contain remains of vertebrates and molluscan shells.
Middle flow (1200 m thick)	Numerous thick ash beds, practically devoid of inter-trappeans sedimentary layers. Exposed in Malwa and Madhya Pradesh.
Lower flow	Associated with numerous fossiliferous inter-trappean beds and rare ash beds. Exposed in parts of M.P. and Eastern areas.

———UNCONFORMITY———

Lameta formation/Bagh beds/Jabalpur formation/Older rocks.

West (1958) recorded the occurrence of 48 flows, five beds of gglomerate, about 280 metres thick inter and intra-trappeans beds and 120 metres of alluvium on the basis of three borings in the Deccan plateau of Eastern Saurashtra and Ahmedabad districts.

Age of the Deccan Traps. The enormous areal extent and thickness of Deccan traps and the presence of inter-trappean beds in between the consecutive lava flows are sufficient to indicate that the formation as a whole has a considerable range in the geological time-scale.

The age of the Deccan Trap is fixed by

(i) the presence of fossils chiefly plants found in the inter-traps,

(ii) the age of the underlying beds (called the infratraps),

(iii) the age of the overlying beds.

(a) In the Narmada valley the traps are underlain by the Bagh beds of upper cretaceous age, possibly in part equivalent to the lametas.

(b) In Surat-Broach there is said to be a distinct erosional unconformity between the top of the traps and the Nummulitic strata for the basal-eocene contains materials derived from the denudation of the traps.

(c) In Kutch, the traps overlie unconformably the Jurassic and lower cretaceous beds, and are overlain by the Numulitics. Here also there seems to be an unconformity between the traps and the numulities though this is not very clear.

(d) In Sind, in the Bor hills near Ranikot, flows of contemporaneous basalt have been described as occurring below and above the carditabeaumonti bed. This bed also contains a nautilus, some corals, echinoids, and gastropode which appear to indicate an upper cretaceous age.

(e) Recent work on the intertrappean fossils especially by B. Sahni and his collaborators, lands supports a lower eocene age for the beds from which the fossils were obtained.

(f) Smith Woodwords work on the fish remains from the lametas has shown that perhaps they are more allied to eocene than to cretaceous formations.

(g) The infra-trappean beds exposed at different places have provided indirect evidences for assigning ages to the Deccan traps, which lie above them. These beds at Dudukuru in West Godavari district have yielded several forms of gastropods, lamellibranchs and nautilus. The fauna from this locality has some affinities with the forms recorded from the cretaceous succession of Trichinopoly district.

(h) Some work has been done on the radio-active character of 7he Deccan traps by V.S. Dubey and R.N. Sukheswala. The work of the latter seems to show that the traps range in age from upper cretaceous to perhaps as late as oligocene.

All the evidences indicate that the Deccan Traps are above 0 million years.

CHAPTER 89

TERTIARY GROUP

During the Tertiary era, there occurred significant changes in the physiographic, environmental, faunal and floral characteristics of the earth. The rise of the Himalayas which took place in a series of five remarkable phases of uplift. The ferns, cycads, conifers which were dominant during the Mesozoic period were replaced by the angiosperms. The reptiles and the ammonites of the mesozoic time were practically extinct. Besides, the marine environment which was existing in the extra-peninsular region, gradually become estuarine and after the principal phase of Himalayan upheaval it changed over to fresh-water environment.

The tertiary rock, in India, are well developed in the extra-peninsular region. In the peninsula small patches of Tertiary rocks occur in Rajasthan and along the coastal tracts in Orissa, Madras, Gujarat etc.

All along the length of the Himalayas tertiary rocks occur in the foot-hill region. Tertiary rocks ranging in age from middle-miocene to middle pleistocene are together known as Siwalik system which are particularly well developed in the western part of the Himalayan Region.

The tertiary succession in the Himalayan region are as follows :

Siwalik System	Upper	Boulder Conglomerate	Low. to mid. pleistocene
		Pinjor	Up. pliocene to
		Tatrot	Low. pleistocene
	Middle	Dhok Pathan	Low. pliocene
		Nagri	Up. miocene
	Lower	Chinji	Mid. miocene
		Kamlial	Mid. miocene

	Upper	
Muree Series	Lower	Lower to Middle Miocene
	Fatehjang zone	

——————UNCONFORMITY——————

Chharat stage   Mid. to Up. Eocene.

Hill limestone   Low. to Mid. Eocene.

Points to remember :

1.   The Chharat stage and Hill limestone are well developed in the Potwar plateau. In Simla-Garhwal area both of them are absent and, in stead there occurs laterites, shales, sandstones and impure limestones which together known as Subathu beds.

2.   The Muree series consists of three zones as (*a*) Fatehjang zone (*b*) Upper and Lower Murees. These rocks contain a few remains of pelecypods and some leaf impressions of angiospermous plants (Sabal major).

3.   The equivalents of Muree series in the Simla-Garhwal area are known as Dagsahi a̓nd Kasauli beds.

SIWALIK SYSTEM

By the time that the lower miocene period was coming to a close, the second phase of the Himalayan Orogeny had commenced. The floor of the Tethys was so much upheaved that the sea had disappeared leaving no basin anywhere even in the flanking areas of Sind and Burma. As the mountains were getting imperceptibly higher and higher, the numerous streams that had already established their courses became more and more active by an increase in their gradient. Thus being rejuvenated, they scoured their valleys with increased vigour and the enormous quantities of the products of decudation were deposited at the foot of the mountains. There also appears to have existed an extensive longitudinal depression in front of the rising mountains and it was in this depression that most of the detrital matter was laid down. The basin behaved as a geosyncline and it subsided as more and more sediments were laid therein. This fluviatile detritus was later subjected to mountain building forces during the third phase of upheaval of the Himalayas and gave rise to a series of hills called Shivailk hills. Thus the Siwailk system named after the Shivalik hills is developed all along the foot hills of the Himalayas Arc.

The system is composed mainly of sandstone, conglomerate, siits and clays which were found under brackish and fresh water environment. An appreciable portion of the Siwalik strata is devoid of fossils. Still then the mammalian fossils are of paramount significance in dividing this system into various sub-divisions.

Climate. It has been inferred that the Kamlial, Chinji and Nagri stages were laid down under a warm and humid climate while the Dhok Pathan period witnessed a relatively dry climatic condition. The upper Siwaliks, on the other hand appear to have been formed under a cold climate.

The abundance of the remains testifies to the very favourable conditions of climate, hydrology, food and suitable environment for entombment of the remains. This system is of highest biological interest as it encloses a rich collection of petrified remains of animals of the vertebrate sub-kingdom not far distant in age from our own times and according to the universally accepted 'doctrine of descent' the immediate ancestors of most of our modern species of land mammalia.

Fossil Assemblages :

Low Siwalik. Primate-Sivapithecus, Brahmapithecus, Dryopithecus.

Carnivora. Vishnufelis, Amphicyon, Dissopsalis.

Proboscidae. Dinotherium, Trilophodon, Steglophodon.

Suidae. Listriodon, Sanitherium.

Equidae. Hipparion.

Rhinoceratidae. Gaindatherium, Aceratherium.

Girraffidae. Girraffokeryx, Giraffa.

Middle-Siwalik : Primate. Maccus, Sivapithecus.

Carnivora. Lutra, Amphicyon, Felis.

Proboscidae. Trilophodon, Tetralophodon, Steglophodon.

Suidae. Sus, Listriodon, Hyosus.

Equidae. Hipparion.

Giraffidae. Vishnutherium, Brahmatherium, Giraffa, Hydaspitherium.

In the Nagri stage—No proboscidae is found.

Upper Siwalik :

Primate. Papio, Simia, Semnopithecus.

Carnivora. Lutra, Panthera, Crocuta, Felis.

Proboscidae. Pentalophodon, Stegodon, Steglophodon.

Equidae. Equus sivalensis (one toed horse).

Suidae. Hyppohyus, Sus, Tetracaidon.

Besides the above, remains of certain birds, reptiles (crocodiles, lizards, turtles, and snakes) and fishes also occur in the Siwalik system. Special interest attaches to the occurrence of about eleven genera of fossil primates in the Siwalik-group.

CHAPTER 90

TERTIARY OF ASSAM

The tertiary succession of rocks are well-developed in the North-Eastern and South-Eastern parts of Assam, where they exhibit a more less complete geological succession ranging in age from palaeocene to lower-pleistocene.

STRATIGRAPHIC SUCCESSION

Series	Stage	Age	Remarks
Dihing series	—	Low pleistocene to pliocene	Developed in all areas
———UNCONFORMITY———			
Dupi Tila series	—	Pliocene to miocene	In Up. Assam known as Namsang stage
———UNCONFORMITY———			
Tipam series	Girujan clays	Low miocene	Developed in all series
	Tipam sandstone	Low miocene	
Surma series	Bokabil stage	Low miocene	Developed in all areas but thin in Up. Assam
	Bhuban stage	Low miocene to Up. oligocene	
———UNCONFORMITY———			
Barail series	Renji stage	Up. oligocene	In Up. Assam Tilak Parbat Bargaloi stage Naogaon stage
	Jenam stage	Low oligocene	
	Laisong stage	Up. eocene	

5

Jaintia series	⌈ Kopili stage	Up. eocene	⌉ In Up. Assam
	│ Sylhet stage	Mid. to low. eocene	│ the oldest tertiary rocks are known as Disang
	⌊ Therria stage	Palaeocene	⌋ series

The Disang series is composed of unfossiliferous shales and sandstones and in the lower and Central Assam, the Jaintia series is composed of three stages. The therria stage is composed of limestones and sandstones, the sylhet stage is composed of limestone beds with intervening horizones of sandstones. The lower sandstone horizons contains seams of coal. The Kopili stage is composed of alternating horizons of shales and sandstones.

The Barail series is composed of three stages and are mostly sandstones, shales, carbonaceous shales and coal seams. In addition to coal, deposits of petroleum are associated with Barails. The Nahorkatiya field derives oil from the Barails.

The Surma series is composed of sandstones, arenaceous shales, conglomerates, etc. The Bhuban stage of the Surma series is said to contain a few oil-bearing horizons, one of which was worked in the Badarpur oil-field.

The Tipam series is divided into two stages, of which the lower Tipam sandstones are oil-bearing and the oil fields at Digboi are of Tipam sandstone age. The Girujan clays is made up mainly of molted clays with some interstratified beds of sandstone, sandy clay and lignite.

In central and lower Assam, the Tipam series is overlain by the Dupi Tila stage, which is composed of sandstones and clays, which are devoid of fossils.

The Tipam and Dihing series are together considered equivalent to the Siwalik system of North-Western Himalayas.

Fossil contents. The remains of foraminifera which are found to occur in the rocks of the tertiary system of Assam include— Nummulities, Discocyclina, Alveolina, Assilina, etc.

CHAPTER 91

PALAEONTOLOGY

FOSSILS

The word 'fossil' has been derived from the Latin word 'Fossilium', which literally means anything dug out of the earth. These are recognisable remains of once-living plants or animals, most of which have been extinct for many thousands of years. They were preserved in sediments, rocks and other materials such as ice, tar, amber etc. prior to historic times. Thus, the remnants of plants or animals of the past geologic ages preserved in the rocks of the earth's crust by natural processes are known as 'fossils'.

Nature and mode of preservation. The varieties in the fossils correspond to what is preserved and how it is preserved. Thus they are of the following nature :

(*i*) Fossils may comprise the remains of the complete animal but that is very rare. Such fossils chiefly include insects preserved in amber, animals in asphalt, and mammoths and other mammals frozen in ice. But such fossils are of rare occurrence and are of recent origin.

(*ii*) There may be petrified remains of hardparts of body in rocks. These are often incomplete and are found in broken fragments.

(*iii*) There may be just the impressions of footprints or leafprints and not the original part of the organism.

(*iv*) Fossils might be also in the form of casts or mould.

Conditions favourable for preservation. We know that millions of animals and plants had lived, died and were destroyed without leaving a trace. But it has been observed that two factors are favourable for the preservation of organisms as fossils

(*a*) the possession of hard parts such as shells and bones, and

(*b*) quick burials of the remains by different processes to prevent destruction by scavengers and decay.

Any animal or plant satisfying the above two conditions can be preserved as fossils under normal conditions. The condition in which fossils occur depends on their original composition and on the material in which they are embedded.

Fossilisation may occur in several ways. Sometimes the soft parts remain unaltered in fossilisation ; sometimes only hard parts remain unaltered and sometimes the hard parts are also altered.

1. Unaltered-soft-parts (Actual remains). In such cases the whole of the organism, including its soft parts is preserved as such. It may be possible due to entombment of the animals under a thick cover of ice. Sometimes insects become entangled in soft and sticky secretions (resin) of trees. On exposure, this hardens and changes to amber and with it the entangled are perfectly preserved.

2. Unaltred hard parts. Shells and internal skeletons are frequently preserved for long periods of time. Many of the best of these are fossils of marine animals that fell into the soft sediment on the sea-floor when they died. Of the land dwellers, those that live near swamps, lakes or sea are most likely to be preserved. Corals, mollusca and protozoans are examples of this category.

3. Altered hard parts. In this case, the actual remains of an organism are likely to undergo changes through time. These changes are fostered by the slowly circulating ground waters that carry elements in solution. It includes the following processes :

(*a*) **Petrifaction.** It is a slow process which involves removal in solution, of each individual molecule of the material constituting the hard parts and simultaneous precipitation of an equivalent quantity of the replacing mineral. This molecule by molecule replacement of one substance by another helps in preserving even the most delicate organic structures, as such. In this manner the bones, shells or plant tissues are transformed into calcite, silica or pyrite and the processes are known as calcification, silicification and pyritization respectively.

(*b*) **Carbonisation.** In this process, the organism is decomposed and it loses nitrogen, oxygen and other volatile constituents. As a result, it is enriched in carbon and is said to have been carbonised. Coal seams are the carbonised remains of plants.

(*c*) **Moulds and casts.** Sometimes, the hard parts preserved within the accumulating sediments, may be totally removed in solution. As a result, hollows are left within the rock beds which are called moulds. Porous and permeable rock-beds are able to have seepage of sub-surface water which can remove in solution the shells of organisms, thus facilitate the formation of moulds.

When the moulds are filled up subsequently with mineral matter, it is known as 'cast', Thus casts only retain the external form of the hard parts.

(*d*) **Impression.** Plants and animals devoid of hard parts, do sometimes leave a record of their existence, in the form of imprints within the rock beds.   Impression of leaves, feathers of extinct birds are the examples.

(4) **Tracks and trails.** While moving on soft and damp ground, the foot prints or trail of the animals are entombed in the mud and when this ground hardens into a rock the foot prints present are preserved.   Even though they do not form any part of the animal, yet they are regarded as fossils.

Uses of fossils :

1. The study of fossils provides evidences in favour of organic evolution and migration of plants and animals through ages.

2. They help in establishing the geoloigcal age of rock beds and their correct order of succession in any area.

3. In the reconstruction of palaeogeography of earth, fossils are paramount significance.   For example, presence of fossil trees or stumps indicate the terrestrial environment and in a similar way the fossils of corals and echinoderms, etc. are suggestive of marine environment.

4. Fossils also help in ascertaining the palaeoclimate.

5. In correlating rock-beds of one area with those of another.

6. The relics of ancient life can be used to establish the time sequence of sedimentary rocks.

7. Fossils provide evidences of "Ontogeny recapitulates phyllogeny'.   Ontogeny means the development of the individual, Recapitulation means brings and phyllogeny means the race history.   This is obvious as the fossils include the ancestors of modern forms.

8. The study of fossils constitutes an integral part of any investigation leading to the discovery of new deposits of coal and petroleum.

9. The study of fossils animals and plants is of the highest importance to the biologist not only because they include the ancestors of modern species but because among fossil forms we find many groups which are altogether exitnct and which often throws light on the relationships of existing animals and plants.

BRACHIOPODA

The animals belonging to the phylum 'Brachiopoda' are characterised by the presence of a *shell* or *test* composed of two valves, which are of unequal nature to each other. The ventral valve is larger than the dorsal valve and bears a circular opening. The ventral valve is also known as *pedicle valve*. The dorsal valve is known as *brachial valve*, also.

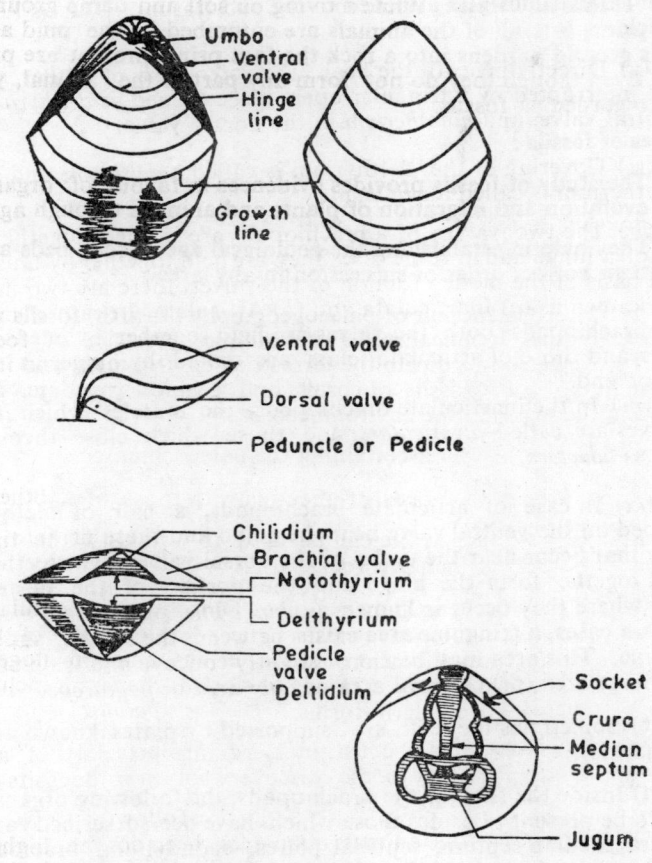

Fig. 91·1. Brachiopoda.

The important features of a brachiopod shell are as follows :

(*i*) The shell is equilateral and unequal.

(*ii*) Both the valves are produced, into what is called *beak* or *umbo*, each. They mark the posterior end of the shell. The umbo of the ventral valve is more prominent than that of the dorsal valve.

(*iii*) The surface of the shell is nearly smooth, but at irregular intervals it is marked by concentric growth lines, which are parallel to shell margin.

(*iv*) Through the foramen of the ventral valve, a muscular stalk which extrudes gets attached to the substratum. This is known as pedicle.

(*v*) Directly beneath the beak, the inter-area of either valve may be interrupted by a tringular open space termed as *delthyrium* in the ventral valve, and *nototherium* in the dorsal valve.

(*vi*) Covering of the delthyrium is called a *deltidium* and that of nototherium is called a *chilidium*.

(*vii*) The two valves of a brachiopod are joined together by means of a hinge and in some cases they are held together by muscles. On the basis of the mode of union of the valves, there are two major classes known as (*a*) Inarticulata and (*b*) Articulata. In case of inarticulata brachiopods both the valves are held together by means of muscles and those of articulata class are united by hinge, at their posterior end.

(*viii*) In the inarticulate brachiopods, the muscles which open the valves are called *divaricators* and those which close them are known as *aducters*.

(*ix*) In case of articulate brachiopods, a pair of teeth are developed on the ventral valve near the umbo and these fit into two sockets that occur near the umbo of the dorsal valve. The teeth and sockets together form the hinge and the portion of the posterior margin where they occur is known as *hinge-line* or *cardinal-margin*. In certain cases, a tringular area exists between the umbo and the hinge-line. This area may be flat or slightly concave and is described as the hinge-area (or) cardinal area or *inter-area* or *palintrope*.

(*x*) Sometimes the teeth are supported by plates known as the dental plate.

(*xi*) Inside the shell of the brachiopods, the following organs are found to be present ; besides those which have been described earlier, *viz.* crura, median septum, septilial plates, spondylium, brachidium which consists of a pair of simply curved on doubly bent arms known as loops and the skeletal connections joining both the arms are known as jugum.

(*xii*) On the external surface of the valves semetimes ridges are found which extend radically from beak. Sometimes there occurs radial ribs, spines, knobs or tubercles, which ornament the external surface of the shell.

The ridges on the shell of the brachiopods are known as *plica-*

tions or *costae* and shells possessing them are described as plicate or costate.

(*xiii*) The shell form of the brachiopods may be elongated, ellipsoidal, sub-circular, alate (winged), concavo-convex, biconvex, plano-convex, resupinate etc.

(*xiv*) When the shells are found to possess minute holes, they are known as punctate shell.

(*xv*) The shells are mainly calcareous in composition, and are formed of three layers :

(*a*) Inner calcareous layer.

(*b*) Middle layer of flattened prisms of calcite.

(*c*) Outer part consists of chitinous material.

(*xvi*) The junction between the valves is known as commissure.

(*xvii*) The initially formed shells are called 'protegulum'. In old age, the shells become thick and their margins truncated. The ornamentation tends to disappear.

Classification :

Phyllum Brachiopoda

Class. Inarticulata (Gastrocaulia)	Articulata (Pygocaulia)
Order :	
(*i*) *Atremata*	(*i*) *Palaeotremata*
Superfamily	*Superfamily-Rustellacea*
(*a*) Lingulacea	
(*b*) Trimerellacea.	
(*ii*) *Neotremata :*	(*ii*) *Protremata and Telotremata :*
Superfamily :	**Superfamily :**
(*a*) Acrotretacea	(*a*) Kutorginacea
(*b*) Siphonotretacea	(*b*) Orthacea
(*c*) Discinacea	(*c*) Productacea
(*d*) Craniacea	(*d*) Pentameracea
	(*e*) Terebratulacea
	(*f*) Spiriferacea
	(*g*) Atrypacea.

Geological range. They are confined to mostly marine environment, hence are found to be present in rocks of marine-nature.

Lower cambrian.   Lingulella, kutorgina, obolella etc.

The geological ranges of some important species are as follows :

1. **Productus.**   Carboniferous to permian.
2. **Orthis.**   Ordovician to silurian.
3. **Spirifer.**   Silurian to permian.
4. **Atrypa.**   Ordovician to devonian.
5. **Rhynchonella.**   Jurassic.
6. **Terebratula.**   Eocene to pliocene.
7. **Terebratella.**   Triassic to present day.

The brachiopods reached the acme of their development during the ordovician, silurian and devonian periods. Within the rocks of tertiary age, brachiopods occur only occasionally.

PELECYPODA (Lamellibranchs)

Pelecypods belong to the phylum 'Mollusca'. The lamellibranchs are oysters. All are marine, some live on land, others in water and many on both. The body is bilaterally symmetrical. The shell consists of two valves which are equivalves and are placed on the right and left side of the body. The valves are inequilateral, *i.e.*, a perpendicular line drawn from the umbo to the opposite margin does not divide the valve into two equal halves.

Both the valves are hinged together at their dorsal margin by means of teeth, sockets and with ligaments. Each valve has an umbo and near it the hinge-line, which marks the dorsal region of the animal. The region where the valves separate most widely when the shell opens is the ventral region. The margin near the mouth is anterior.

Sometimes there is infront of the umbones an oval-shaped depressed area of smaller size appears like a groove, shared by both the valves. It is known as *'lunule'*. Behind the umbones, there is a similar somewhat larger area, known as *'escutcheon'*.

The hinge is formed by projections known as teeth alternating with depressions or grooves known as sockets. The teeth and sockets in a hinge line. alternate with each other in the two halves and teeth of one valve fits into the sockets of the other valve.

Depending on the nature, shape and size of the teeth and sockets, a few distinct types of dentition have so far been recognised. The dentitions are as follows :

(*i*) **Taxodont.** Also known as etenodont. Here the teeth and sockets are more or less similar in form and size, *e.g.*, Nucula, Arca.

(*ii*) **Dysodont.** Here the teeth radiate outwards form the umbo, *e.g.,* Mytilus, Ostrea.

(*iii*) **Isodont.** Here the hinge consists of two strong and slightly curved teeth, which occur in each valve and fits into the sockets of other valve, *e.g.,* Spondylus.

(*iv*) **Schizodont.** Sometimes a few thick and occasionally grooved teeth are developed and these may vary in shape and size.

(*v*) **Heterodont.** Also known as telodont, where the teeth are not of uniform shape and size and are few in number.

(*vi*) **Desmodont.** Here true teeth and a hinge plate, but one or more laminae or ridge are developed at the hinge margin. Commonly one cardinal tooth in each valve, *e.g.,* Pleuromya.

(*vii*) **Asthenodont.** In the burrowing lamellibranchs, the hinge is made up of obsolete teeth and the dentition is described as asthenodont.

(*viii*) **Edentulous.** Lamellibranchs with no teeth are said to possess edentulous shell. This is also known as palaeoconcha.

The surface of the shell may be smooth or may be ornamented with radiating ribs, concentric ribs and striations, tubercles and spines etc. The margins of the valves may be smooth or crenulated.

The interior of the valves are marked with the impressions of the muscles. In a living animal there are elastic ligaments which perform the function of divaricators in case of lamellibranchs.

Usually two adductor impressions occur in the interior of each valves. One of these two is placed anteriorly and the other posteriorly. Both the impressions are connected together by means of a linear depression called the pallial line, which runs more or less parallel to the ventral margin of the valve. Sometimes there is a notch on the pallial line known as the *pallial sinus*.

When two adductors impressions are found to occur the shell is called a '*Dimyarian shell*'. When there is only one impression the shell is said to be a *monomyarian shell*. When both the impressions are of equal size, it is known as *Isomyarian* shell.

When the ligament lies below the hinge-line, it is called a *resilium*. A process for lodging the resilium is called *chondrophore*. Where the ligament extends on either side of the umbo, it is called amphidetic ligament but when it is entirely behind the umbo, it is described as Opisthodetic ligament.

Byssus is a thread like process derived from anterior portion of foot and used to attach the shell to substratum.

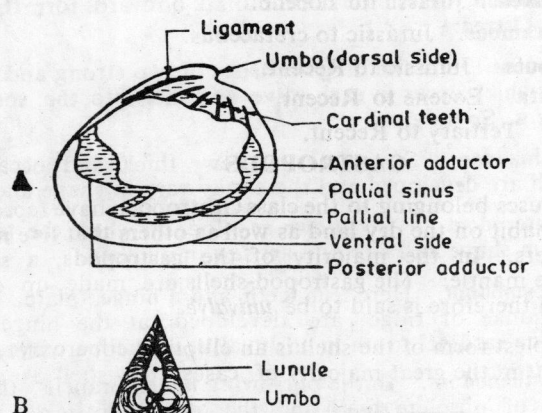

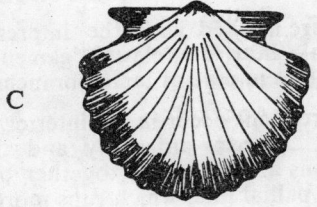

A—Interior of left valve.   B—Dorsal view of a shell.   C—Pecten.
Fig. 91·2. Lamellibranch.

The first formed embryonic shell is called the 'Prodissoconch'.
The shape of the pelecypod shells may be alate, rostrate, mytiliform,
quadrate, produced etc.

Geologic history. The earliest records of fossil lamellibranchs
have been traced from the rocks of lower-ordovician age, and they
reached their culmination during the *quarternary age.* The geological
range of some important pelecypods are as follows :

    (*a*) **Arca.** Jurrassic to Recent.

    (*b*) **Nucula.** Silurian to Recent.

    (*c*) **Pecten.** Carboniferous to Recent.

    (*d*) **Ostrea.** Triassic to Recent.

(*e*) **Gryphaea.** Jurassic to Eocene.

(*f*) **Inoceramous.** Jurassic to cretaceous.

(*g*) **Trigonia** Jurassic to Recent.

(*h*) **Cardita.** Eocene to Recent.

(*i*) **Mya.** Tertiary to Recent.

GASTROPODS

The molluscs belonging to the class Gastropoda have representatives which inhabit on the dry land as well as others that live in fresh and salt waters. In the majority of the gastropods, a shell is secreted by the mantle. The gastropod-shells are made up of one valve only and therefore is said to be *univalve.*

The simplest form of the shell is an elliptical cone widely open at the base, but in the great majority of cases, the shell is a cone

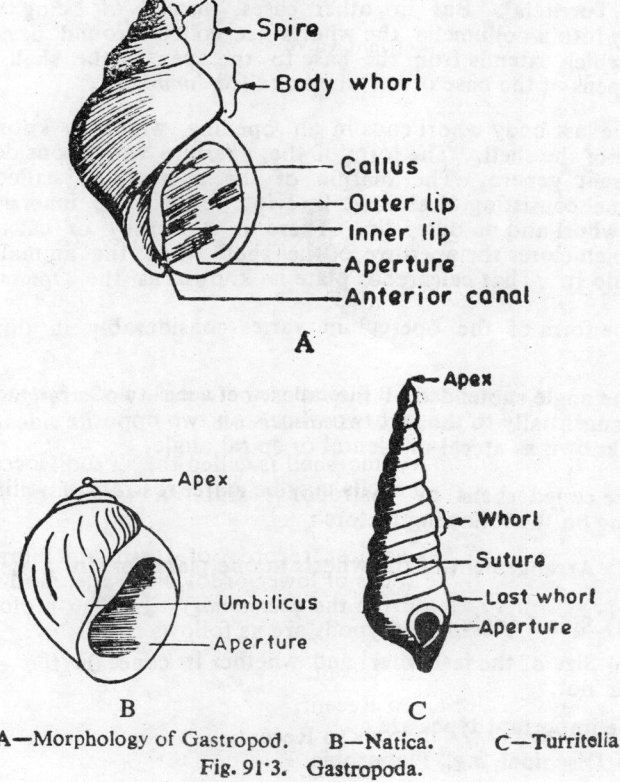

A—Morphology of Gastropod.   B—Natica.   C—Turritelia.
Fig. 91·3. Gastropoda.

coiled into a helicoid spiral, open at one end and tapering to a point at the other. Each one of the individual coils is described as a *whorl*.

All the whorls except the last, together form the *spire* of the shell. The tapering end of the spire is known as '*apex*' and it forms the posterior end of the shell. The part of the whorl farthest from the apex is called the base of the shell.

In a few genera, the whorls are separated, but as a rule they are in contact. The successive whorls lie in mutual contact along a line called the '*suture*'.

The coiling of the shell is usually dextral, *i.e.*, when the apex of the shell is pointed upwards, the aperture will be on the right hand side, *i.e.*, (Clockwise). But in a few cases it is sinstral (anticlockwise) when the aperture occurs on the left hand side, *e g.*, is 'Physa'.

Sometimes the inner faces of the whorls are united into a solid pillar extending from the base to the apex and is known as *columella*, *e.g.*, is 'Turritela'. But in other cases, instead of being united centrally into a collumella, the whorls are coiled around a central cavity, which extends from the base to the apex of the shell. The cavity opens at the base of the shell ; called '*umblicus*'.

The last body whorl ends in an opening, which is known as aperture of the shell. The form of the aperture varies considerably in different genera. The margin of the aperture is called thf 'peristome' consisting of an inner lip lying towards the innerside oe the last whorl and an outer lip. There is a horney or calcareous plate which closes the aperture of the shell when the animal withdraws into it. That calcareous plate is known as the 'Operculum'.

The form of the operculum varies considerably in different genera.

The angle subtended at the apex between two straight lines drawn tangentially to the last two whorls on two opposite sides of the shell, is known as apical or pleural or spiral angle.

The coiled shells of gastropods exhibit a variety of forms, depending on the following factors :

(*i*) Arrangement of the whorls in one plane or in a helicoid spiral.

(*ii*) Spiral angle.

(*iii*) Size of the last whorl and whether it conceals the earlier whorls or not.

The important types are :

(*a*) Discoidal, *e.g.*, Planorobis.

(*b*) Conical or Trochiform, *e.g.*, Trochus.

(*c*) Turbinate (*i.e.*, resembling a spinning top), *e.g.*, Turbo.

(*d*) Turreted (where spire is long and shape is that of an acute cone), *e.g.*, Turritella.

(*e*) Fusiform (where the last whorl repeats the shape of the spire in inverted position), *e.g.*, Fusinous, Rimellarimosa.

(*f*) Cylindrical, *e.g.*, Pupilla.

(*g*) Globular (last whorl forms the greater part of the external surface, the spire being very low), *e.g.*, Natica.

(*h*) Convolute (hemi-ellipsoidal, no spiral coiling is recognizable), *e.g.*, Cypraea.

(*i*) Auriform – here the spire is short and the aperture is conspicuously large.

The surface of the shell is frequently ornamented with spines, knobs, ribs or striae. In some genera like Murex, rows of spines, or lamellar processes, extend across all the whorls from the apex to the base of the shell, forming what are termed as varices, these are elevations.

The embryonic shell is often found at the apex of the shell and is called the protoconch.

Classification of gastropods. The zoological classifications of gastropods are usually accepted, which takes into account the nature of their soft parts.

The scheme of classification of gastropods are as follows :

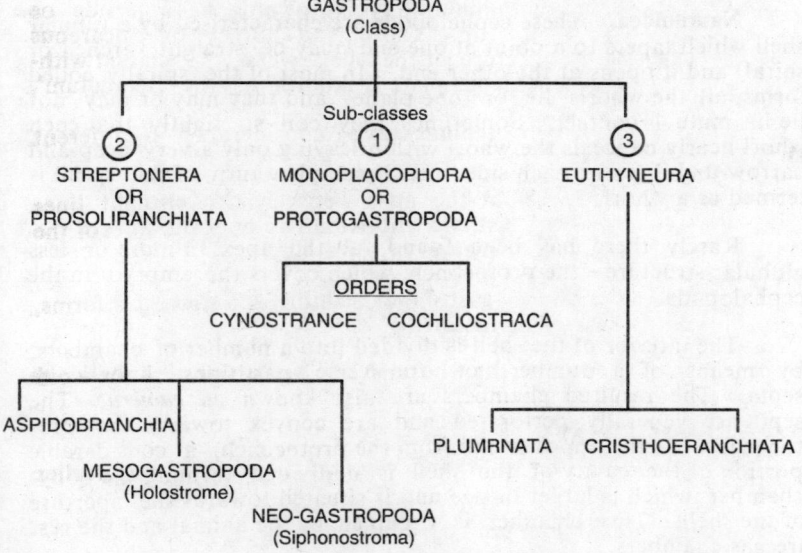

Geological Ranges :

(*i*)	Trochus	Triassic to present day.
(*ii*)	Nerita	Cretaceous to present day.
(*iii*)	Natica	Triassic to present day.
(*iv*)	Turritella	Cretaceous to present day.
(*v*)	Cerithium	Cretaceous to present day.
(*vi*)	Buccinum	Pliocene to present day.
(*vii*)	Murex	Eocene to present day.
(*viii*)	Voluta	Eocene to present day.
(*ix*)	Conus	Up. cretaceous to present day.
(*x*)	Planorobis	Jurassic to present day.

CEPHALOPODA

The cephalopods are entirely marine and are highly organised than other molluscs. Nearly all cephalopods retain perfect bilateral symmetry and are covered by exoskeleton. On the basis of morphology, the class cephalopoda has been divided into three major sub-classes as :

1. Nautiloidea, 2. Ammonoidea, 3. Dibranchia.

The 1st and 2nd classes are sometimes grouped together as *tetrabranchia*.

Nautiloidea. These cephalopods are characterised by a tubular shell which tapers to a point at one end (may be straight, arched or spiral) and it opens at the other end. In most of the spirally coiled forms, all the whorls lie on one plane, and they may or may not lie in mutual contact. Sometimes they coil so tightly that each whorl nearly conceals the whorl within leaving only a very deep and narrow umbilicus on each side. Each complete turn of the spiral is termed as a whorl.

Rarely there has been found, at the apex, a more or less globular structure – the protoconch, which covers the embryo in the cephalopoda.

The interior of the shell is divided into a number of chambers, by means of a number of transverse partitions known as septa. The resulted chambers are also known as *camerae*. The septa are generally perforated and are convex towards the protoconch. At the opposite end from the protoconch, a considerable portion of the cavity of the shell is undivided forming the body-chamber which is larger in size and is situated towards the aperture of the shell. This chamber is occupied by the animal and the rest are gas chambers.

Through the central perforation of the septa ran a living cord, the siphuncle, through all the gas-chambers back to the protoconch. The calcareous tube is made up of two parts known as the *septal neck* and the *connecting-ring*.

While the animal grows, new shell is continually being added to the aperture or mouth of the body chamber ; while at intervals the animals move forwards in its body chamber and secrete behind it a new septum.

The line of junction of the edge of a septum with the external shell is termed the *suture-line* or *septal-sature*.

A straight conical shell of a nautiloid is known as *orthoceracone* and slightly curved shell is described as *cyrotocone*. One of the chief characters of the shell in the nautiloidea is the *simple form of the suture-lines*.

Ammonoidea. In the ammonoid-cephalopods the shell is generally coiled into a plane spiral, and as a rule the suture-lines show a complicated pattern. The body of the typical ammonoids varies from less than half a whorl to over two whorls in length. The lateral regions seen within the encircling last whorl is called the umbilicus. The spiral line of contact between each whorl and the next is the umbilical suture.

The portions of the suture-lines which are convex-towards the mouth of the shell are called saddles and the intervening concave portions are known as lobes. The saddles and lobes are nearly always similar on either side.

There are three types of suture lines in different types of ammonoids as

(*i*) **Goniatitic suture-line**. These suture lines have pointed lobes and with rounded saddles.

(*ii*) **Ceratitic suture-line**. These suture-lines are found to consist of cerrated lobes and rounded saddles. *e.g.*, ceratite.

(*iii*) **Ammonitic suture-line**. These are complicated types of suture line and are found to consist of cerrated lobes and cerrated saddles. *e.g.*, ammonites.

Basing on the form of suture line, the ammonoids have been classified as goniatite, ceratite and ammonites.

The surface of the ammonoids-shell may by smooth or may be ornamented with striations, ribs, tubercles or spines. In some ammonoids at the internal margin of the shell, there is a ridge known as *keel*.

A straight conical shell of an ammonoid cephaloid is known as *bactriticone*. The following features distinguishes an ammonoid from a nautiloid :

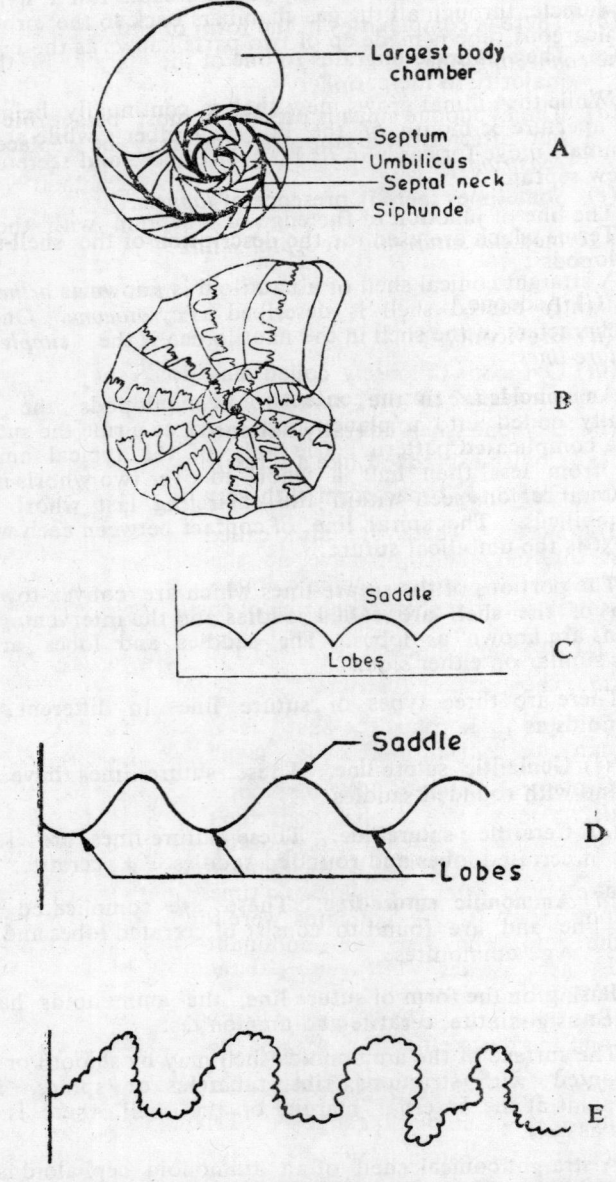

Fig. 91·4. A—Cephalopod, B—Ceratite suture, C—Goniatite suture,
D—Simple suture line, E—Ammonitic suture line.

(*a*) The protoconch is usually barrel-shaped.

(*b*) There is complexities in the form of suture line.

(*c*) The siphuncle migrates to one of the margins in the great majority to the periphery.

(*d*) The hypnomic sinus is present in most palaeozoic genera, but afterwards disappears, and may be replaced by a *roastrum*.

(*e*) Sometimes there is presence of *lappets*.

Terms which are used for the description of the shell-shape of cephalopods :

(*i*) Asocone,

(*ii*) Brevicone,

(*iii*) Gyrocone (a loosely coiled nautiloid shell in which the whorls are not in contact),

(*iv*) Oxycone (shell laterally flattened),

(*v*) Conispiral (coiled like a screw),

(*vi*) Cadicone (broad form with very deep umbilicus).

Dibranchia. These are also called coleoidea. They either possess internal shells or have no shells at all. The well known extinct genus 'Belemnites', which belongs to the subclass dibranchia, possesses an internal shell consisting of three parts :

(*i*) The phragmocone   (*ii*) Pro-ostracum

(*iii*) The guard or rostrum.

Of these three parts, the guard is readily preserved as fossil. The guard is a conical, more or less cigar shaped solid.

Geological history. The cephalopods are exclusively marine animals. Among the cephalopods the nautiloids are said to be the most ancient forms ; even found to occur in rocks cf lower cambrian age. They attained their maximum during permian and carboniferous. After the palaeozoic times they suddenly decreased in importance. The ammonoids were found in the Devonian times and the goniatites were the dominant forms. During triassic period the ceratites were quite abundant and ammonites were abundant during cretaceous.

But the dibranchia made their first appearance in the mesozoic times and the representative form was 'belemnite'.

Goniatites	Geological range	Carboniferous
Ceratites	Geological range	Triassic]
Phyloceras	Geological range	Jurassic to cretaceous
Lytoceras	Geological range	Jurassic to cretaceous

Acanthoceras	Geological range	Jurassic to cretaceous
Peltoceras	Geological range	Jurassic to cretaceous
Schloenbachia	Geological range	Jurassic to cretaceous
Baculites	Geological range	Upper cretaceous
Turrilites	Geological range	Upper cretaceous
Belemnites	Geological range	Upper cretaceous

CORALS

Corals are benthonic colenterata, which commonly occur in the form of colonies. Corals belong to the class 'Anthozoa'. The polyps are the individual animals. The entire skeleton of a simple coral is described as the corallum. In case of compound corals, the skeleton of each individual member of a coloney is known as corallite.

The body of a coral has a more or less cylindrical or conical shape. The base of the cone is generally depressed and is known as the calyx. The calcareous wall which forms the boundary of the corallum is called '*thecca*' and an outer wall called '*epitheca*'.

According to the nature and internal structure of the corals, there are five subclasses as

(a) Zoantharia   (b) Alcyonaria

(c) Rugosa   (d) Tabulata

(e) Schizocorallia.

Corals are sometimes extremely abundant and built up extensive reefs and banks, which are known as *bioherms.*

Corals are known to have existed even during the early palaeozoic era. Geologic age of some important corals are as follows :

(*i*) *Calceola.* Middle devonian.

(*ii*) *Zaphrentis.* Devonian to carboniferous.

(*iii*) *Favosites.* Carboniferous.

(*iv*) *Halysites.* Palaeozoic era.

TRILOBITES

These animals belong to the sub-class Trilobita, class-crustacea, of Phylum—Arthropoda.

In all trilobites, the body is divided into three longitudinal lobes as well as three transverse portions. This trilobation of the body gives the name 'trilobita'.

In trilobites, the body is flattened from above downwards and is divided into three parts by means of two furrows, which extends from the anterior to the posterior extremities. The body of the trilobite is made up of three distinct parts known as (*a*) the head (*b*) the thorax (*c*) the pigidium. Of the three lobes of a trilobite, the lateral ones are known as *pleural lobes*, while the intervening central lobe is described as *axial lobes*. Each part of the body is segmented. The dorsal surface of the body is protected by a strong calcareous exoskeleton, known as the dorsal shield.

The part of the dorsal shield, which covers the head of the animal, is known as the cephalic shield or head shield or cephalon.

(*a*) **Cephalic shield.** It is separated into three parts by a median and two lateral portions. The medial portion is more convexed than the lateral portion and is called the *glabella*. The lateral portions are called cheeks or genae. By means of two axial furrows, the glabella and cheeks are separated from each other.

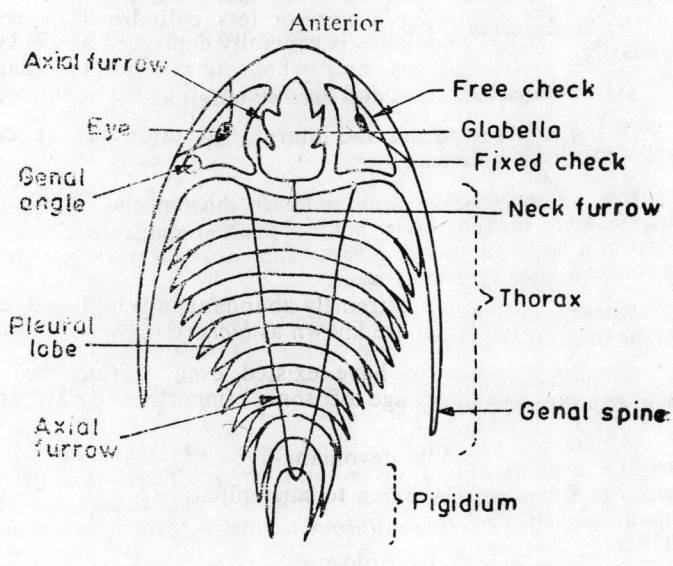

Fig. 91·5. Morphology of trilobita.

Glabella. It is segmented. The last segment of the glabella, lying farthest from the anterior margin of the cephalon is known as the occipital lobe or neck lobe. The occipital lobe is separated from the rest of the glabella by an occipital furrow.

Cheeks. The posterior angle of the cheek is known as the genal angle, which is subtended between the lateral and posterior margins of cephalon. Sometimes the genal angles are produced posteriorly in the form of spines, which are known as genal spines. The facial suture divides the cheek into two portions ; the inner part of the cheek between the glabella and the facial suture is called the fixed cheek. The portion of the cheek lying external to the suture is known as free cheek or librigena.

The glabella and the two fixed cheeks together form what is called cranidium. According to the course of facial suture, trilobites are of the following types :

(*i*) **Opisthoparian**. When the facial suture commences on the posterior boarder inside the genal angle towards the anterior boarder.

(*ii*) **Proparian**. Facial suture leaves the head shield via the lateral margin.

(*iii*) **Gonatoparian**. Facial suture cuts genal angle.

(*iv*) **Hypoparian**. Facial suture-marginal and developed secondarily.

(*v*) **Protoparian**. Facial suture is marginal.

Owing to the fusion of the fixed cheek and free cheek, the facial suture is sometimes absent.

Eye. The eyes are compound eyes, located one on each free cheek located in the angle made by the facial suture. The eyes consist of a large number of lenses which are biconvex or globular. In a few trilobites eyes are absent.

Sometimes there is a protuberance of the fixed cheek fitting into the outline of the eye, called the *palpebral lobes*.

Sometimes a thread like ridge called the eye-line found to extend from the eye to the glabella.

(*b*) **Thorax**. It consists of a series of segments separated from the cephalic-shield by the neck-rings. The part of the dorsal shield, which lies posterior to the cephalon and covers the thorax, is made up of two to forty-two segments, which are not fused together.

Each thoracic segment is divided into three lobes by two axial furrows. The central lobe is known as the *axial-lobe* and the lateral ones are called *pleurae*. A bit off from the axial lobe, each pleurae is found to be curved and is known as *fulcrum*. Sometimes the lateral portions of pleurae are flat plate-like, which are called facets.

(*c*) **Pygidium**. All the hind end is a tail-piece showing very well marked segmentation. This is known as pygidium. It is commonly triangular or semicircular in shape. Here all the segments are fused together. Similar to thorax it is divided into axial and pleural portions. The part of the dorsal shield which covers the pygidium is known as the caudal shield or abdomen. It is made u p of 2 to 29 trilobated segments.

Sometimes the posterior-end of the caudal shield is provided with a spine, which is described as '*telson*'.

The possession of antennae and the biramous character of the other appendages connect the trilobites with the crustacea.

When the cephalon is larger than the pygidium, the trilobites are said to be *heteropygous* ; when they are approximately of the same size, they are termed as *Isopygous*.

Geological range. Trilobites occur as fossils only within the rocks of palaeozoic ages. They suddenly appeared in the lower-cambrian and reached their maximum in cambrian and ordovician period. In silurian, they were still abundant but became less important in devonian. With the end of the palaeozoic era, they became extinct. Geological age of some important species :

(*i*)	*Olenellus*.	Low-cambrian.
(*ii*)	*Paradoxides*.	Mid-cambrian.
(*iii*)	*Olenus*.	Up-cambrian.
(*iv*)	*Calymene*.	Ordovician.
(*v*)	*Illaenus*.	Ordovician.
(*vi*)	*Phacops*.	Silurian.
(*vii*)	*Redlichia*.	Mid-cambrian.
(*viii*)	*Phillipsia*.	Carboniferous to permian.
(*ix*)	*Asaphus*.	Ordovician.

In view of their wide geographical distribution and rather a limited range in geological time, trilobites are regarded as excellent index-fossils.

ECHINOIDS

The echinoderms, belonging to the class Echinoidea, are also known as sea-urchins. They always possess a compact and rigid, more or less a globular, heart shaped or discoidal body covered with spines. The shell or test is covered by a layer of ectoderm and consists of numerous calcareous plates which constitute the exoskeleton. Some of the echinoids show radial symmetry and others show bilateral symmetry.

The lower side of the test is known as the *oral or ventral side*. The upper side of the test is generally convex in shape and is termed as the *dorsal* or *aboral* side.

The mouth is situated on the oral side, either in a central position or in front of the centre. The anus is either at the summit of the test or posterior to it. The mouth of the animal is encircled by a number of plates which collectively constitute what is known as the peristome. In the same way the anus is surrounded by a number of plates, which together constitute what is called the *periproct*.

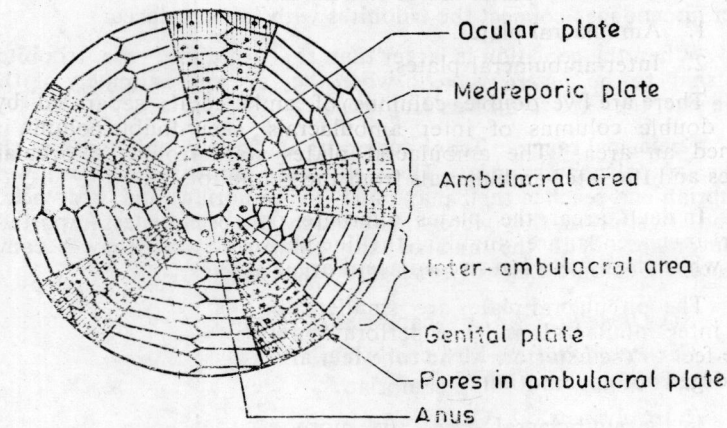

Fig. 91·6. **Aboral view of echinoid.**

Sometimes both the mouth and anus are placed at opposite poles of the test and in some other echinoids the anus is always and the mouth is often excentric. Accordingly, the echinoids are divided into two orders like

    (*i*) Regularia   (*ii*) Irregularia.

The test or shell of an echinoid is made up of three parts :

    (*a*) Apical system or apical disc,

    (*b*) Corona,

    (*c*) Peristome.

Apical disc. Around the anus there is a flat series of ten plates called the apical-disc. Of the ten plates, five are known as *occular plates* and the rest are called the *genital plates*. The genital plates are larger than the occular plates and they lie in contact with the periproct and form the inner part of the ring. One of the genital plates is perforated by a great number, of smaller pores, through which water filters into the water-vessels : this plate is known as the *Medreporic plate*. All the genital plates are interradial and are termed as *basals*.

The occular plates lie in alternation with the genital plates and constitute the outer part of the ring. They are termed radials. When the occulars separate the genitals and touch the periproct, they are said to be *insert* and when they do not touch the periproct they are *exsert*.

(*b*) **Corona.** It is the main part of the test made up of twenty columns of plates ; each column extending from the apical-disc to the peristome. The plates are of two kinds :

1. Ambulacral plates.
2. Inter-ambulacral plates.

There are five double columns of ambulacrals separated by five double columns of inter ambulacrals ; each double column is termed an area. The ambulacral plates start from the occular plates and the inter-ambulacrals from the genital plates.

In each area, the plates alternate on either side and since their inner ends are angular, the line between the two rows is zig-zag, while those between two adjacent areas are straight.

The ambulacral plates are smaller and more numerous than the interambulacral and are perforated by pores for the passage of tube-feet to the exterior. The tube-feet are called podia, which are the organs of locomotion.

In the umbulacral area, the more aboral plates are called simple-primaries which have one pore-pair in each plate. Lower-down, compound plates appear. Each pair of pore is surrounded by an oval raised rim, known as *peripordium*. Compound plates also consists of various types of plates known as demiplates, occluded plates, included plates etc.

Echinoids are often provided with rounded elevations known as granules or tubercles. They are of various sizes. The larger ones are called *pirmary* and the smaller ones are called the *secondary*.

Parts of a tubercle :

(*i*) At the top of the elevation, there is a hemispherical piece with a perforation (at the top) known as *mamelon*.

(*ii*) The mamelon rests on the *boss*. The base of the boss is generally surrounded by a smooth excavated groove known as *areola* or *scrobicule*, to which the muscles of the spines are attached.

The spines consist of the following parts :

(*i*) **Acetabulum.** It is that part of the spine which fits to the mamelon.

(*ii*) **Collar.** Next to the acetabulum is the head of the spine above which lies a ring called 'collar'.

(*iii*) Beyond the collar and forming the greater part of the spine is the *shaft* or *stem*, which may be smooth or ornamented with ridges or rows of spiny processes.

Sometimes very small tubercles occur in the form of bands. Such bends of minute tubercles form what are known as *fasciole*.

(*c*) **Peristome.** It lies on the ventral side of the test and the mouth of the animal occurs at the centre of the peristome. The peristome is made up of plates which are loose and are either isolated or occur in pair. The position of mouth in the peristome also varies in different forms. Peristome is usually a memberane that surrounds the mouth.

Geological range. It has been recorded that echinoids made their first appearance during the palaeozoic time, particularly in *ordovician*, with regular echinoids.

(*a*) *Cidaris*. Jurassic to present day. (Regularia).

(*b*) *Hemicidaris*. Cretaceous. (Regularia).

(*c*) *Echinus*. Pliocene to present day. (Regularia).

(*d*) *Pygaster*. Triassic to cretaceous. (Irregularia).

(*e*) *Echinolampas*. Lower eocene to present day. (Irregularia).

(*f*) *Holaster*. Tertiary to present day. (Irregularia).

(*g*) *Micraster*. Triassic to tertiary. (Irregularia).

(*h*) *Hemiaster*. Cretaceous to present day. (Irregularia).

GONDWANA FLORA

The Gondwana time for India, ranges from the Upper-Carboniferous to Lower-Cretaceous. The gymnosperms and pteridophytes were the predominant flora in the Gondwana time. These were vascular plants.

Plumsted (Asutralia) from the study of carboniferous and permian flora suggested that there were three distinct types of flora during this time. They were :

(*i*) Lepidodendron flora   Oldest.

(*ii*) Rhacopteris flora   Low-carboniferous.

(*iii*) Glossopteris flora   Youngest.

In India and other southern landmasses of southern hemisphere, the Gondwana rocks are characterised by the presence of Glossopteris flora. This flora is said to have been evolved from those plants which could survive the permocarboniferous glaciation.

According to D. N. Wadia, there is a correspondence of the three types of Gondwana flora with that of the three sudivisions of Gondwanaland. According to him (*a*) **Glossopteris flora**. Low. Gondwana (Lower-Triassic)

(*b*) **Dicroidium (Thinnfeldia)**. Mid-Gondwana (Lower-Jurassic).

(*c*) **Ptilophyllum flora**. Up. Gondwana (Low-Cretaceous).

Glossopteris flora. They inculde plant-fossils like glassopteris, gangamopteris, vertebraria, schizoneura, noeggerathiopsis, gondwanidium, buridia phyllotheca etc.

The glossopteris, gangamopteris, vartebraria are all pteridosperms. They are leaf genera.

Glossopteris is represented by fronds. (A leaf which bears fructification is a frond). Here mid-rib is well marked and there is reticulate venation. Leaf is contracted at the base.

Gangamopteris is similar to glossopteris but is without any mid-rib.

Vertebraria is thought to be the stem of glossoptris (knows as rhizome).

The important plant are as follows :

1. **Pteridosperms**. Glossopteris, gangamopteris, vertebraria.
2. **Corditales**. Noeggerathiopsis (leaf), dadoxylon (stem).
3. **Equisetales**. Schizoneura, phyllotheca.
4. **Sphenophyllales**. Sphenophyllum.
5. **Cycadophyta**. Taeniopteris.

(*b*) **Dicroidium (Mid-Gondwana flora)**. The common flora is thinnfeldia. It occurs abundantly in the northern continents and dicroidium in southern continents. Thinnfeldia flora indicates lower jurassic age and Dicroidium flora indicates lower triassic age.

(*c*) **Ptilophyllum flora**. It is dominated by the cycads and the filicales. Here the pteridosperms are of no importance, while the conifers appear to have attained some amount of development.

Cycadophytes. Bennettitales, Ptillophyllum, Pterophyllum, Dictyozomites, Otozamites (all leaves) Williamsonia (flower), Bucklandia (stems).

Fern. Cladophlebis, Marattiopsis, Sphenopteris.

Conifers. Brachyphyllum, Elatocladus, conites.

Equisetales. Equisetitias.

Lycopodiales. Lycopodites.

The gymnosperms and the cycads of the mesozoic era forming the Rajmahal flora,disappeared gradually and during the cretaceous flowering plants began to appear on the surface of the globe.

The contemporary flora of the world during the Gondwana time are as :

(*i*) **The Eur-American Flora.** Best known in U.S.A., Europe, Iran, Turkey, and contain lepidodendron, neuropteris, pecopteris, sigillaria erc.

(*ii*) **The Cathaysian flora.** The Gigantopteris flora extended to Korea and North China, and southwards into Indochina, Siam, sumatra and it also occupied north-America and Taxas.

(*iii*) **The Angaraland flora.** From Ural mountains eastwards to Pacific and from north of Korea to Bering Strait and to south it extended to Mongolia and almost down to Pamear Region.

These flora are contemporaneous with glossopteris flora.